VALLEY

OF THE

APES

The Search for Sasquatch in Area X

Michael Mayes

ANOMALIST BOOKS
Charlottesville, Virginia

An Original Publication of Anomalist Books
Valley of the Apes: The Search for Sasquatch in Area X
Copyright © 2022 by Michael Mayes
ISBN: 978-1-949501-22-3

Cover image: Photo taken by Michael Mayes in the Sam Houston National
Forest in August 2005.
Back cover image: Ouachita Mountains, Shawn Herrington/Shutterstock

Book design: Seale Studios

Anomalist Books
3445 Seminole Trail #247
Charlottesville, VA 22911

CONTENTS

For Alton Higgins

FOREWORD

The events detailed in this book provide some of the best reasons to be hopeful that a resolution to the Sasquatch mystery is close at hand. What you're about to read is a meticulously documented collection of observations made by people—academics, wildlife professionals, and scientists, among others—whom I consider to be among the most reliable individuals to have ever pursued the Sasquatch. Operating under the formal rubric of the North American Wood Ape Conservancy (NAWAC), the organization has maintained cohesion for well over a decade in an effort that requires the cooperation and collaboration of dozens of individuals sacrificing time, energy, and resources year after year to maintain a constant vigil in an unforgiving wilderness.

The man telling the story, Michael Mayes, has been at the forefront of this endeavor since its inception. I first met Mike in 2017 at an NAWAC gathering near its primary study site (Area X) in southeastern Oklahoma. Mike is an educator and a coach by profession, and it didn't take much time spent with him for me to see that he's optimally wired for those roles. An exemplary teacher and qualified coach, Mike achieves the dual goals of providing meaningful motivation while defining expectations for the highest quality performance from both himself and his colleagues.

I'm very grateful that Mike has taken on the task of telling you this story. I think he's done a remarkable job of compiling years of information into a single narrative. The importance of this story might not be fully appreciated until the Sasquatch is officially recognized, but it is a story that needs to be told in the interim.

There is a great deal that we don't know about Sasquatch phenomenon. What I do know after having spent the last several years with the NAWAC leads me to make this prediction: When the day comes that people say "They proved that the Sasquatch exists," you, the reader, probably won't have to wonder who "they" are. You will have already met them right here in the pages of this book.

Matt Pruitt
Producer of *Bigfoot and Beyond*
with Cliff and Bobo
February 2022

AUTHOR'S NOTE

The problem with truth is that it is hard to believe. It is even harder to get other people to believe. — Walter Darby Bannard

This is not your father's Bigfoot book. What you are about to read is not another breakdown of the Patterson-Gimlin film, or a retelling of classic Sasquatch tales like the sighting at Ruby Creek, the Albert Ostman abduction, or the Ape Canyon incident. This is different.

If you have ever wondered what Bigfoot research is *truly* like, and whether anyone was making a serious effort to solve the wood ape conundrum, then this is the book for you. Here, you will learn what it is really like to pursue the most elusive animal on the continent across some of the most inhospitable terrain imaginable. You will hear what it is like to endure blistering heat, subfreezing temperatures, and massive thunderstorms while out in the field. You will read about the risks of venturing into remote wilderness locations and encountering poachers, venomous snakes, wild hogs, black bears, and the occasional mountain lion. You will learn how difficult this endeavor actually is and, perhaps, come away with a better understanding of what researchers are up against in their efforts to document this species.

You will also come to better understand how human foibles have interfered with the effort to secure incontrovertible evidence that this creature exists. From time to time, you will likely find yourself asking, *Why did they do that? Why didn't they do this?* or *Why didn't they do X instead of Y?* The answer to these questions is: because we are human. As my friend and fellow researcher Matt Pruitt often points out, there is no generational knowledge about how to hunt these animals, no "tricks of the trade" that have been handed down through the years, and no how-to manual. Lack of familiarity with the terrain, illness, sleep deprivation, and, at times, flat out fear, have all led to mistakes. We have tried to learn from these mistakes and endeavor to not repeat them. Mike Tyson once said, "Everyone has a plan...until they get punched in the face." We have been punched in the face—figuratively—many times and kicked ourselves about missed opportunities. One thing we have done, and will continue to do, is pick ourselves up off the canvas and keep trying.

What you are about to read is all true. Every word. For many, some of the events documented in this book will seem too incredible, too fantastic, to have occurred. I assure you that they did. You have not only my word on that but the word of every member of the North American Wood Ape Conservancy (NAWAC). I suppose that you will have to decide what that is worth. I will not be spending any additional time or energy trying to convince you to believe me or my fellow NAWAC members. You will believe us or you won't.

I joined the NAWAC—then the Texas Bigfoot Research Center (TBRC)—in the fall of 2005. I approached the group with some trepidation. While I had spoken with a member who seemed sane enough, I had read up on a lot of Bigfoot research groups online and was extremely disappointed with what I found there. With very few exceptions, little in the way of actual scientific research was being undertaken. But my concerns about the men and women of the TBRC were quickly allayed after attending a public meeting. After watching Alton Higgins, a college professor and wildlife biologist, give a fascinating presentation on the stride length of the Patterson-Gimlin film subject, I was sold. I was truly in the company of people who were using scientific methodology in an attempt to crack this mystery. I joined the group immediately upon the conclusion of the meeting. I have never regretted that decision.

Through the NAWAC I have experienced some incredible things, things that I have not shared with most of my friends or family. I knew that they would not believe me, or worse, think I was a liar or a crackpot. I know that most of my NAWAC brethren are as close-mouthed about the subject as I am, and for the same reasons.

After all these years, I feel it is now time to open up and share our theories, methodology, experiences, and what we have learned about an animal that is not supposed to exist outside of the realm of myth.

Welcome to the valley of the apes.

Michael Mayes
Temple, Texas
January 3, 2022

NOTE: Names followed by an asterisk (*) on first mention are pseudonyms. In most cases, these names represent former or non-members whom I was unable to reach in order to gain the permission necessary to use their real names in this book. In a few cases, individuals—for reasons of their own—simply preferred I not use their real names. In those instances, I have chosen to honor their request and protect their privacy.

PREFACE

Daryl Colyer forced himself to move slowly and methodically despite the quickly sinking sun. Alex Diaz trailed behind him, maintaining an approximate distance of 25 yards between himself and his teammate. Another team member, Travis Lawrence, had quickly moved to the southwest in an effort to get to the creek bed. Once there, he would bend back to the northeast. This strategy, the men hoped, would allow them to catch their target in a pincer movement.

Just moments before, the men had heard a loud bang on the metal roof of the west cabin. The sound had been far too loud to have been caused by a falling hickory nut. The group was sure it had been a rock impact, a rock thrown by an ape. Experience told the men that the thrower of the rock would likely stay in the area, hoping for a reaction from the group.

Colyer slowly advanced to the spot where the trail bent gently back to the northwest. He had continued to hear movement in the woods ahead and believed it was likely attributable to the rock-thrower. He was determined to see who, or what, was responsible for the loud impact he and the other men had just heard.

After rounding the bend in the trail, Colyer caught sight of his quarry. The figure was large, walked upright, and had a conical-shaped head covered in brown hair; it was an ape. The creature was no more than 90 feet away but was quickly moving away to the south. It was now or never. Colyer took aim at the ape, which was now entering the dense woods, with his semi-automatic 12-gauge shotgun and emptied all nine rounds in its direction.

Diaz rushed to Colyer's side from around the bend in the trail. Bluish-white smoke hung in the air for several seconds compromising the ability of the men to see in the direction of their target. As the smoke began to clear, the men hurried to the spot where, moments before, an animal thought to be only a myth by most had been traversing.

What they saw when they got there was something neither man would ever forget.

INTRODUCTION

It would be the height of arrogance for me to assume that everyone is familiar with the North American Wood Ape Conservancy (NAWAC). Therefore, an introduction is in order. The NAWAC, as currently organized, has been in existence since 2013, but its roots go all the way back to the summer of 1999, when the Texas Bigfoot Research Center was formed. The organization used that moniker until 2007 when its leaders decided to make the group a 501(c)(3), tax-exempt, non-profit scientific-research organization, recognized by the Internal Revenue Service. When the official paperwork was filed, the name was changed to the Texas Bigfoot Research Conservancy (TBRC) in order to better reflect the goals of the group—not to mention the fact that there had never been a center anywhere in the Lone Star State.

The membership rebranded the organization again in 2013, and the North American Wood Ape Conservancy (NAWAC) was born. We hoped that the new name would help scientifically-minded individuals feel more comfortable about showing a serious interest in the work of the group and cast the entire subject in a more favorable light.

The NAWAC is an all-volunteer group of men and women from many different walks of life. The diverse roster of investigators includes biologists, teachers, doctors, nurses, musicians, writers, naturalists, accountants, entrepreneurs, and other professionals. In addition, many of the investigators are former military, law enforcement, and/or lifetime outdoorsmen with unique skill sets to offer.

The NAWAC is of the belief that the centuries-old "wild man" or "hairy man" phenomenon in North America can be attributed to the existence of an as yet undiscovered great ape. Everything the group does is in an attempt to test this hypothesis.

The NAWAC is funded exclusively by membership dues, fundraisers, donations, grants, and conferences. Every dollar taken in by the group is applied toward the goal of documenting what we believe is an undiscovered North American ape. No one in the NAWAC draws a salary or is reimbursed for labor, travel, or personal expenses. It is truly an all-volunteer organization. The group is governed by a membership-elected Board of Directors comprised of seven people who seek only to pursue the group's mission statement:

To investigate and conduct research regarding the existence of the unlisted primate species we refer to as the wood ape, also known as the sasquatch or Bigfoot; to facilitate scientific, official, and governmental recognition, conservation, and protection of the species and its habitat; and to help further factual education and understanding to the public regarding the species, with the focus on the continent of North America.

Now, allow me to share our story with you: what we have seen, what we have heard, our successes, and our failures. The telling of this story is, I believe, long overdue. Let's begin.

PART 1:
The Early Years

"A mysterious creature—that's news, Ishmael. The fact that people see it—that's news."

— David Guterson,
Snow Falling on Cedars

1
Apes in the South?

For many, Bigfoot was "born" in the summer of 1958. It was in late August of that year that tractor operator Gerald "Jerry" Crew discovered huge footprints on a work site in northern California.[1] Some members of the work crew were members of the Hoopa Valley Tribe, and informed Jerry that the tracks had to have been made by the hairy forest giants the tribe believed inhabited the region. Crew made plaster casts of the giant prints and took them to Andrew Genzoli, the editor of the *Humboldt Times*, a local newspaper. Genzoli ran the story along with a photograph of Crew holding one of the plaster casts on October 5, 1958.[2] In the article, Genzoli referred to the creator of the footprints as "Bigfoot." The story was picked up nationally, and the rest is history.

The possibility that a hair-covered biped really did inhabit the wilderness of North America was only strengthened nine years later, when Roger Patterson and Bob Gimlin captured film footage of a large, hair-covered creature walking briskly across a sandbar at Bluff Creek, California. The 1967 footage has become iconic, and despite what some claim, has never been successfully debunked. The attention drummed up by the Jerry Crew incident and the Patterson-Gimlin film led to what many consider to have been a golden age of Sasquatch research. The efforts were led by the "Four Horsemen": John Green, Peter Byrne, Rene Dahinden, and anthropologist Grover Krantz. Many other serious-minded individuals took a good long look at the phenomenon as well. Opinions varied, of course, but it was encouraging that qualified scientists were not afraid to look into the matter. For example, anthropologist John Napier, who at the time was Director of Primate Studies at the Smithsonian Institution, gave his opinion on the topic in his 1972 paper *Bigfoot: The Yeti and Sasquatch in Myth and Reality*. In reference to the footprints found at the Bluff Creek worksite, he wrote, "They were very big tracks deeply impressed in the soil and had the look of human footprints. They measured 16 inches long and 7 inches wide and they were all over the place. They went up hill and down dale and continued into situations that seemed to defy the ingenuity of a hoaxer with a footprint machine."[3]

But it is important to note that the origin of the footprints found by Jerry Crew is hotly debated. Claims of hoaxing by Crew's boss, contractor Ray

Wallace, have been around for many years, and the sentiment that the prints found on the Bluff Creek jobsite were his doing has picked up considerable traction over time with some researchers. Whether the Bluff Creek prints found by Jerry Crew were legitimate or hoaxed is not really the point here. What is important is that the publicity given to the incident propelled Bigfoot into the public consciousness. Also, important to note is that esteemed professional scientists of the day, like primatologist John Napier, were not afraid of committing career suicide by examining purported Sasquatch evidence and rendering an opinion on it.

While Bigfoot captured the public imagination in a way few things had before, it was considered by most to be only a Pacific Northwest phenomenon. Northern California, Oregon, Washington, and Canadian British Columbia were considered Sasquatch territory. So why would NAWAC members waste their time searching for a creature that—even if real—lived more than 1,000 miles away from their center of operations? While the Pacific Northwest may indeed be the cradle of all things Sasquatch, the assumption that it is the only part of the continent where hair-covered bipedal creatures have been reported is incorrect. Sightings of Bigfoot-like creatures have been reported from coast-to-coast over the last two centuries. But if the oral histories of Native American tribes are considered, the sightings extend even further back in time. Many such encounters took place in what is now Texas, Louisiana, Oklahoma, and Arkansas—the region which has become the NAWAC's main area of study–and were chronicled by the press. It is true that while the term Bigfoot was never used in these old articles, descriptors such as "wild man," "gorilla," and the like *were* used in an effort to describe what had been seen. In some areas, sightings occurred frequently enough that the creatures were given nicknames. "The Noxie Monster," "the Honka," "Momo," "The Fouke Monster," "Ol' Mossyback," "the Route Monster," "the Chambers Creek Monster," and "the Raggedy Man" are just a few such monikers attached to upright, hair-covered beasts that haunted specific southern communities over a prolonged stretch of time.

Skeptics might claim that residents of the Texas, Louisiana, Oklahoma, Arkansas region jumped on the Bigfoot bandwagon after the Patterson-Gimlin film was made public in the late 1960s. But in truth there is no shortage of historical anecdotal accounts of encounters with creatures that match the modern description of the Sasquatch, predating both the Crew footprint find and the Patterson-Gimlin footage.

In Texas, there were numerous reports of forest-dwelling, ape-like creatures that appeared in print prior to the 1950s and 1960s. Maybe the earliest such example is the tale of the Wild Woman of the Navidad. In 1837, shortly after Texas had won independence from Mexico, settlers along the banks of the Navidad River began to lose crops and goods to a petty thief. The thief, nicknamed "that thing that comes" by the settlers, raided the potato fields, corn cribs, and even entered cabins to steal bread, saws, forks, pots, and pans. These items were precious and hard to come by on the Texas frontier. This went on for years until the locals had finally had enough and organized a plan to capture the culprit. A number of hunters formed an extended line and drove through the woods with leashed hounds where the wild woman was said to reside. Other men, mounted on horseback, took positions outside the wood line in hopes of roping the woman once she had been forced out of the woods and onto the open prairie. It took several tries, but eventually the men succeeded in flushing out their quarry. To their surprise, what sprang from the woods was not a woman–at least not a human woman–rather, it was an incredibly fast and completely hair-covered creature. It managed to evade the lassos of the men due in no small part to the fact that their horses shied away from the creature whenever they drew close—and escape.[4] In 1850, a runaway slave was captured who admitted to the burglaries that had plagued the settlers over the years. For many, the capture of the runaway slave closed the book on the tale of the Wild Woman of the Navidad. The runaway slave was certainly responsible for the thefts attributed to "the thing that comes," but what are we to make of the hair-covered creature who was flushed from the woods and was fast enough to outrun men on horseback? Was it a wood ape?

Another possible example of a historic Texas wood ape sighting appeared in the *Argus*, a Michigan newspaper, on September 1, 1871. The article chronicled the excitement of the people in the Gatesville community "over the appearance of an immense orang outang [sic] in its vicinity." The animal was "described as being about seven feet high and covered from head to foot with a thick coating of hair."[5]

The Mexia *Weekly Herald* of April 7, 1933, detailed the plans of Sheriff Rusk Roane to hunt down the "wild man of the thickets" in an isolated section of Fort Bend County, near De Walt.[6] Sheriff Roane likened the wild man to "Tarzan of the Apes," as he was repeatedly described as having a long beard and hairy body devoid of clothing.

While reports of creatures matching the traditional description of the Sasquatch have come in from all over the Lone Star State, no area can match the volume of sightings that have originated in east Texas. The eastern third of the state is heavily wooded, and water sources are numerous. This is exactly the opposite of what might be expected by many who know Texas only as the dusty and dry desert so often portrayed on the silver screen. In total, Texas has a stunning 62.4 million acres of forests and woodlands,[7] which include vast tracts of hardwood river bottoms that produce prodigious amounts of acorns, hickory nuts, black walnuts, and pecans. This environment supports all manner of wildlife, including white-tailed deer, feral hogs, and small mammals. Noted wildlife biologist John Bindernagel visited southeast Texas and southwest Louisiana in the early 2000s and was impressed with the richness and scope of the mixed deciduous forests found in the region. When compared to the conifer-dominated forests of the Pacific Northwest, Bindernagel asserted that any large species of mammal living in southern forests would almost certainly require a smaller home range than an animal of the same species living in the northwest.[8]

As with sightings from other parts of the state, many east Texas encounters were reported well before the term Bigfoot was coined in the late 1950s. For example, on August 14, 1952, the *Kountze News* ran a story about "wild man sightings" in the Big Thicket. Multiple eyewitnesses were interviewed, and all agreed that the wild man "had a heavy beard and a hairy body," and was always naked and barefoot. Local authorities took the sightings seriously enough to investigate several of these incidents. A few of these investigations yielded physical evidence. In a *Kountze News* article from 1952, the Hardin County Sheriff said, "the barefoot tracks were plain to see."[9]

A woman named Ann Bazan related the experience of her father, an oil company maintenance man, that took place in the 1920s. In those days, there were few, if any, roads running through the dense southeast Texas woods and swamps. Her father, who had been charged with the upkeep of the oil pipelines snaking through this wilderness, rode a horse up and down the right-of-ways, or *senderos*, that had been cut through the woods to accommodate the pipelines. While on one of his outings, her father claims to have been attacked by something big, strong, and covered in hair. The creature attempted to wrestle the oilman from his horse but was repelled and slunk back into the woods. According to Bazan, the incident bothered her father for the rest of his life, and the general consensus that he had been attacked by a lunatic hermit who made the thicket his home just did not sit right with him. According to the oilman, the eyes and face of his attacker were not quite human, not to mention it was completely naked and covered in thick hair.[10]

East of the Sabine River lies the bayou country of Louisiana. Considered a sportsman's paradise, Louisiana has 14 million acres of forestland.[11] In addition to the forests, its rivers, swamps, marshes, and bayous winding through the state create amazingly productive habitats for all manner of creatures, possibly including wood apes. Early accounts out of Louisiana are hard to come by, but there are a few. One was reported in the Caddo, Louisiana, *Gazette* regarding the story of a man who was attacked and snatched off his horse, bitten, and scratched up by a "wild man" while out riding near the Arkansas-Louisiana border in 1856.[12]

While pre-1950s newspaper accounts of ape-like creatures in Louisiana are scarce, tales of other hair-covered bipeds in Louisiana folklore are easy to find. The Cajuns and Creoles of southern Louisiana have long told tales of the *Loup Garou*, a sort of bayou-based werewolf. Barry Jean Ancelet, an expert on Cajun folklore and a professor at the University of Louisiana at Lafayette, says the tale of the *Loup Garou* is a common legend across French Louisiana.[13] The beast is said to stalk the swamps around Acadiana and greater New Orleans, as well as forests across the Bayou State. The legend of the *Loup Garou* has been used to persuade Cajun children to behave for decades, but is it possible there is more to the story than just campfire tales? The *Loup Garou* is said to be very large, have a human-shaped, hair-covered body, and the head of a wolf with glowing eyes.[14] Sightings of such creatures continue to be reported to the present day. Outside of French Louisiana, an encounter with a huge, hirsute creature would almost universally be classified as a Bigfoot sighting. If the legend of the *Loup Garou* is based in fact, then it is entirely possible that what the people of Louisiana occasionally see in the bayous, swamps, and woods of their state is not a werewolf, but a wood ape.

Perhaps the best example of an alleged Louisiana Bigfoot is the beast said to haunt the Honey Island Swamp. First reported by retired air traffic controller Harlan Ford in August of 1963, the creature was described as bipedal, covered in hair, with yellow eyes.[15] After Ford's death in 1980, a reel of Super 8 film was discovered among his belongings. The film, shot somewhere in the Honey Island Swamp, shows a tall, dark, bipedal figure walking along the shore of the marsh. The authenticity of the film footage has been hotly debated ever since. Supporters claim that the fact Ford never went public with the footage shows he was not a con man or a hoaxer out to make a quick buck. They also point out that the Native Americans who made the Pearl River area their home have long told tales of the *Letiche*: a meat-eating, human-like creature that lived both in the water and on the land.[16] Others, less convinced by the footage, simply theorize that Ford did not think his film was good enough to hold up to scrutiny, so he hid it away. Either way, the idea that the Honey Island Swamp holds a monster has taken hold, and reports continue to come out of the area, further strengthening the case that an undocumented primate of some kind may be inhabiting the Bayou State.

Oklahoma has become a second home to the mostly Texas-based NAWAC. This makes little sense to those who picture Oklahoma as nothing more than an almost treeless open prairie. While this is an accurate description of much of the Sooner State, it is an incomplete picture. The extreme eastern portion of Oklahoma is not only heavily forested, but mountainous. The northeastern portion of Oklahoma is dominated by the Ozark Plateau, which makes up only a small portion of the Ozark Mountains.[17] To the south lie the Ouachita Mountains, which extend east to west for approximately 225 miles. Within the range is the Ouachita National Forest, which is rich in foodstuff for all manner of animals including black bear, deer, turkey, foxes, and numerous other mammalian species. The area receives large amounts of precipitation, which have given rise to numerous streams, creeks, and rivers.[18] In all, Oklahoma contains approximately 12 million acres of forestland.[19]

The eastern portion of the Sooner State also has a long history of encounters with large, hair-covered, bipedal creatures. One of the earliest accounts took place in May of 1849, when a McCurtain County hunter had a run-in with a "strange critter" that matched the description of a hairy primate.[20]

In 1915, a young man named Crum King returned to his home in Nowata County after attending a dance only to find a five-to- six-foot tall, hairy, man-like creature standing by the gate of his property. He reported the creature had a wide chest, was about four feet across, and "stood there with its arms stretched out." King turned and fled.[21]

In 1926, a well-known doctor saw "a thing" running across the road in the lights of his Model T Ford while driving near Goodwater.[22] That same year, two hunters spotted what they described as a "big, black, hairy, ape man" near the Mountain Fork River. The beast ran away when one of the hunters set his dog after it. After about an hour of searching, the two men found the corpse of the dog. It had been "almost torn in two."[23]

In 1956, about 13 miles from Wilberton, several people spotted multiple hairy, long-armed creatures walking near a pond on their property. One of the creatures was also seen "hanging over a neighbor's fence."[24] Many other encounters could be included here, but the point is that there is a long history of sightings of creatures matching the description of the Sasquatch or wood ape in Oklahoma, long before the 1958 Jerry Crew footprint story. Sightings continue to come out of eastern Oklahoma on a regular basis. So much so, in fact, that the NAWAC has shifted its main area of study from Texas to southeastern Oklahoma.

Rounding out the four-state region in which the NAWAC operates most of the time is Arkansas. Arkansas also has a long history of sightings of ape-like animals. Certainly, there is ample habitat to support such a creature. Known as the "Natural State," Arkansas is a land of "…unsurpassed scenery, clear lakes, free-flowing streams, magnificent rivers, meandering bayous, delta bottomlands, forested mountains, and abundant fish and wildlife."[25] The terrain is stunningly beautiful, and one could be forgiven for thinking the entire state is one big forest. While that is not quite true, Arkansas does boast 19 million acres of woods.[26] That equates to approximately 56% of the total area of the state. In addition, much of the forestland is rugged, mountainous, and sparsely populated, making it a perfect place for a furtive species of primate to hide.

By far, the most well-known Bigfoot tales to come out of Arkansas are those associated with the so-called "Fouke Monster." The spate of sightings of a hair-covered, man-like beast in Miller County, near the small town of Fouke, in the 1960s spawned the classic Charles B. Pierce movie *The Legend of Boggy Creek*. The first, and perhaps best-known incident involving the Fouke creature is the one involving a then 14-year-old named Lynn Crabtree. Crabtree was out hunting squirrels when he came face-to face with what he described as a seven-to-eight-foot-tall beast, which was covered in reddish-brown hair about four inches in length. Crabtree described a face obscured by hair with "only a dark brown nose showing."[27] When the creature began walking towards him, the boy fired off three rounds with his 20-gauge shotgun in the direction of the animal before turning and fleeing.

While the Fouke area sightings of the 1960s and early 1970s received massive media attention, similar sightings had been occurring in Arkansas for many decades.

The March 1846 issue of *Scientific American* mentioned that a wild man had been seen in the swamps along the Arkansas and Missouri border. The prints left behind by the wild man measured 22 inches, with toes "as long as a common man's fingers."[28]

In 1851, a wild man of "gigantic stature" was reported by residents of Greene County. Other than being exceptionally large, the wild man was described as having a "…body covered in hair, and the head with long locks that fairly enveloped his neck and shoulders." In addition, the creature was able to run away "…with great speed, leaping from twelve to fourteen feet at a time. His foot-prints [sic] measured thirteen inches each."[29] An article in the *Memphis Inquirer* regarding the wild man said, "This singular creature has long been known traditionally in St. Francis, Greene, and Poinsett Counties," and "So well authenticated have now become the accounts of this creature, that an expedition is organizing in Memphis, by Col. David C. Cross and Dr. Sullivan, to scout for him."[30]

In 1856, a "wild man, seven feet high" was spotted by multiple witnesses in the Mississippi river bottoms.[31] Later that same year, a party tried to hunt down a wild man near Brant Lake. One of the searchers reported seeing a creature with the build of "…a stout, athletic man, about six feet four inches in height, completely covered with hair of a brownish cast about four to six inches long. He was well muscled and ran up the bank with the fleetness of a deer."[32]

An account from the 1860s captures what is described as an ape-like creature and is detailed in the 1941 book *Ozark Country* by Otto Ernest Rayburn. Rayburn recounts that "the Giant of the Hills was often seen in the Arkansas Ouachita Mountains" and that "This seven foot 'wild man' was covered with thick hair and lived in caves or by the Saline River."[33] The locals were afraid of the creature, so they set out to capture it. According to Rayburn, they did just that and locked the beast in the Benton jail. But the townsfolk were unable to hold the creature in the small wooden structure that housed the jail; it quickly broke out and escaped.

Admittedly, the capture of the wood ape-like creature described by Otto Rayburn sounds very similar to other tales of Bigfoot captures from the past, most notable among them might be the 1884 story of "Jacko" familiar to Sasquatch enthusiasts. Even if the account relayed to Otto Rayburn is purely a yarn spun by locals at the time; it still suggests that the people of the Arkansas Ouachita region were well-acquainted with the idea of large, hair-covered, wild men roaming the more lonesome reaches of the mountains and forests. Again, the tale precedes the Jerry Crew track find as well as the Patterson-Gimlin film footage by many years.

Of course, anecdotal accounts like these do not constitute proof that the wood ape actually exists in the four-state region; still, they are valuable tidbits of information that indicate the residents of the area are, and have been, familiar with the idea of such an animal for a very long time. Skeptics might argue that anecdotes such as these have little value, that they have no place in modern science. I would argue, however, that *observation* is woven into the very core of science. Natural history is rife with examples of anecdotal accounts from indigenous peoples who provided guidance for biologists and naturalists seeking out novel or extant relic species in regions around the globe.[34] The NAWAC has used anecdotal sighting reports, both historical and contemporary, to zero in on where the target species is living and to help decide where to spend its limited resources to the maximum efficiency.

The forested areas of Texas, Oklahoma, Louisiana, and Arkansas are larger than some entire western states said to harbor wood apes. Though the woods of the four-state region are more fragmented than forests in the Pacific Northwest, there remain huge stands of pine and hardwoods that are more than capable of supporting all manner of wildlife. If the region can support black bears, which it does in great numbers, there is no reason to think a small population of wood apes could not survive there as well. The sheer amount of suitable habitat in combination with historical and contemporary sightings of large, hair-covered bipeds, are the factors that led the NAWAC to conduct research here. The group believes strongly that the same animal that walks the forests around Bluff Creek, California, the Olympic Peninsula of Washington, the Oregon National Forest, and the deep woods of Canadian British Columbia also inhabits the woods and mountain ranges of Texas, Oklahoma, Louisiana, and Arkansas.

2
Weekend Warriors

The Texas Bigfoot Research Center (TBRC), which later became the NAWAC, was not unlike most Bigfoot research groups when it was first formed. The people in the group were interested in Bigfoot for different reasons. Some members had reported seeing one of these creatures in the past or had heard something in the woods for which they had no explanation; others simply found the entire phenomenon fascinating on an intellectual level and were intensely curious. The one thing all the members had in common was a burning desire to get to the bottom of the mystery: to find the truth. Just how to go about doing that was the question at hand.

As the group studied the history and methods of researchers who had come before, they discovered that almost all efforts to investigate the subject in the past had been reactive in nature. Typically, there would be a Bigfoot sighting somewhere, maybe more than one; the word would get out and the investigators would arrive. Sometimes the researchers would not get to the sighting location until days or weeks after the incident took place. While valuable evidence was sometimes collected in the form of footprint casts, photos of trackways, witness statements, and the like, the creature responsible for the excitement always seemed to have been long gone. This reactive method seemed to have very little chance of producing concrete evidence that would move the wood ape from the dark realm of myth and folklore to the list of documented and cataloged animals of the world.

TBRC leadership decided to take a more proactive approach to investigating the wood ape phenomenon. There would still be investigations of sightings after the fact—information considered valuable to establish where these animals were living in the four-state region—but the organization also wanted to go out and attempt to elicit a response from the animals being sought. "We want to get right in their grill and get them worked up," said Field Operations Coordinator Daryl Colyer, "We want to get a reaction out of them."

So in the early 2000s the group began to organize outings, usually two to three days in duration, for members. Places like Miller County, Arkansas, the Big Thicket National Preserve, the Sam Houston National Forest, and the

Davy Crockett National Forest were chosen, places with a history of sightings of tall, bipedal, and hair-covered creatures. Some members also took vacation trips to Washington, Colorado, Ohio, Minnesota, and California in an attempt to find these apes.

The group insisted on doing things the right way, which sometimes meant seeking the cooperation of governmental agencies and acquiring the necessary permits. Obtaining permission from state or federal agencies to conduct operations on public land in order to gather evidence of the existence of a creature that does not officially exist proved to be a difficult proposition at times. It would have been easier to just proceed with the plans and hope that no game wardens or rangers showed up to ask what the group was doing. That, however, was not the way the TBRC wanted to do business, and so every effort was made to comply with all laws and regulations.

Once a team was on-site, the operation leader would go over the plans and assign responsibilities to individual members. Looking back, the methodology used seems a bit simplistic, but the group was learning as it went along. This was new ground, and the only way to learn was through trial and error. The group would experience plenty of both in the years to come.

The equipment used by members on these operations generally consisted of whatever each individual brought along, which means that the equipment available during these operations varied from one outing to the next. The equipment typically consisted of 35mm cameras, video recorders (most with Night Shot capabilities), Gen I, II, or III night vision monoculars, trail cameras, flashlights, red headlamps, bionic ears, audio recorders, pheromone chips (when available), self-contained sound broadcasting equipment, and Garmin Rino 120 GPS/two-way radio units.

Participants would meet at a predetermined location, usually on a Friday afternoon or evening. Briefings would be held to familiarize participants with the objectives of the operation, the terrain and conditions they could expect to encounter, team assignments (night or day), the placement of teams, and the scheduled times when call-blasts of suspected wood ape vocalizations would be played.

Once briefed, night teams would be ferried out to study areas by vehicle. The vehicles drove slowly, and sound discipline was stressed. Teams were always instructed to whisper, even when communicating by radio, and to remain as silent as possible. Flashlights were used only if absolutely necessary. Once the destinations were reached, members would utilize scent covers and hike out to their assigned areas some distance from the speakers that would be broadcasting the vocalizations. The idea was to reduce the chance of animals associating the broadcast sounds with humans in any way (this is why the organization rarely played calls from base camps). Typically, groups would sit quietly for 60-90 minutes before call-blasting would begin. The researchers hoped that a reply could be recorded or that video footage could be captured of a wood ape sufficiently intrigued to approach the area.

Night operations typically lasted until about 3:00 a.m. If no activity was noted by that time, the teams would usually call it a night and return to base camp. Bright and early the next morning, group members who had been assigned to day teams would comb the areas where call-blasting had taken place the night before in order to look for physical evidence. Teams would search for tracks, hair, scat, suspicious tree breaks, etc. At least one member of the day team would have a video recorder ready at all times in case the target species was sighted. Day teams usually worked an area from 8:00 a.m. to mid-afternoon. The thought was that giving the area a break from human activity for a few hours might put the indigenous wildlife at ease before night operations started again.

As these operations usually took place over a weekend, some group members would typically arrive in time to conduct night activities on Friday, and more would arrive the next morning to participate on the day team. The night teams would go back out Saturday evening and were followed by day teams who did an abbreviated inspection of the study area early Sunday. Generally, members had to leave the study area by midday to return to their homes–some hundreds of miles away–and get ready for work on Monday. The "weekend warrior" nature of these operations limited the amount of real study done on a given area and made the chances of encountering the target species slim, at best.

This procedure was pretty much standard during the early years of the group. Techniques were refined over time, and the group did try to throw some curveballs in an attempt to trick an ape into revealing itself. Despite the obvious limitations of this methodology, the group did have some startling successes while employing these tactics. Some of the more interesting experiences from these early field operations are chronicled in the next few chapters.

3
Operation Primate Web II

Operation Primate Web II took place between January 6 and 9, 2005, in the Sam Houston National Forest (SHNF) of southeast Texas. (Nothing of note occurred during the first Operation Primate Web.) The forest is located approximately 50 miles north of Houston and contains 163,037 acres of land in Montgomery, Walker, and San Jacinto Counties.[35] The SHNF receives approximately 44 inches of rainfall per year and is rich in water sources. The San Jacinto River runs through the region, as do numerous creeks and streams. Lake Conroe is located in the western third of the forest, and Lake Livingston, which is fed by the Trinity River, is to the immediate north and east of the forest proper. The forest itself is a rich combination of pine, mixed, and hardwood bottom ecosystems.

The SHNF was chosen as the site for this operation based on more than a dozen credible reports of large, bipedal primates in the area over the previous 20 years. These sightings, which included two encounters by law enforcement officers, provided the impetus for a large-scale (by group standards) field

Michael Mayes

A large, upright "something" beating a hasty retreat in a remote area of the Sam Houston National Forest not far from where Operation Primate Web II was conducted. I took this photo in August of 2005, a few months after having my first visual in May of that year.

study. The operation was scheduled to begin at the conclusion of deer season.

Operation Primate Web II was unique in that it was a joint effort between members of the TBRC and the Bigfoot Field Researchers Organization (BFRO). Such cooperative efforts between research groups would prove rare in the future. The goal of the operation was simple: recover hard evidence that would lead to the documentation of this enigmatic species in the Sam Houston National Forest. The goal of documentation could be achieved through clear video footage and/or photographs of the target species, identifiable and viable DNA samples of the target species, and/or the acquisition of hair samples, scat samples, foot or handprint castings, or unique sound recordings. At the very least, the investigators hoped to identify areas of habitation and high activity for the target species.[36]

The operation kicked off on Thursday evening, January 6, with the deployment of three teams–comprising a total of eight people–into a remote area of the SHNF in San Jacinto County. Call-blasting began at 7:00 p.m. and continued intermittently until 9:30 p.m. when rain showers threatened to ruin the team's many sensitive electronic devices. Miserably damp and cold conditions–temperatures ranged from the mid 30s to the mid 40s–would plague the operation from beginning to end. Even so, the outing would yield some spectacular results before wrapping up.

Day teams were sent out the next morning to two different areas in search of signs of the target species. The teams found much wildlife spoor but no evidence that might indicate wood apes were in the area. That evening, night teams were deployed in a semi-circular arrangement (the "Primate Web") measuring approximately 0.5 miles in diameter in an area of the SHNF west of Lake Conroe. The call-blasting team was located at the center of the semi-circular "web." The groups also deployed pheromone chips, provided by Greg Bambenek out of Minnesota who, at the time, was a BFRO Curator.[37] The chips were made of plastic and impregnated with a sexual scent attractant similar to that found in gorillas. The chips were extremely pungent and team members were able to smell them up to 40 feet away.

Call-blasting began promptly at 7:00 p.m. The broadcasts seemed to stir up the wildlife in the area, and many vocalizations were detected. Coyotes and barred owls were the most commonly reported animals heard by team members. At 8:00 p.m., they played a broadcast of the "Ohio Howl," a long, moaning, howl-like vocalization, thought by many to have been produced by a wood ape and recorded by BFRO Chairman Matt Moneymaker in 1994.[38] Almost immediately, a return call was heard by all five teams in the field. Everyone agreed that the return vocalization was almost identical to the broad-cast call, though it was slightly higher in pitch. The return howl had come from the southeast, where no vocalizations of common wildlife had been heard that night. The decision was made to continue with the broadcasting.

At 9:22 p.m., one team reported hearing vocalizations similar to those of a "spider monkey."[39] These sounds were followed up by "whoops" and something that sounded similar to a peacock. At 10:40 p.m., a team made up of two Louisiana firefighters positioned near the center of the semi-circular web, reported hearing what sounded like a large animal walking around their post. The sounds continued for the next hour. The men found the sounds a bit disturbing, and not like the behavior of any known animal that should have been in the area, but they held their ground.

After this initial excitement, things slowed down considerably. There were no more return vocalizations, no more animal sounds of any kind, and whatever had been circling the one team of investigators had seemingly left the area. The decision was made to call it a night shortly after 2:00 a.m. As the two firefighters stood up to begin gathering their gear, a thick tree branch approximately two feet long came flying out of the darkness and landed within 10 feet of the stunned pair. Both men felt strongly that the branch had been flying in their direction horizontally based on the fact that it skidded closer to them after making contact with the ground. A falling branch might bounce up after hitting the ground; it would not have the momentum to then skid horizontally in any direction. The two firefighters notified the team leaders and were told to stay put; team members were on the way and would extract the men. TBRC members Daryl Colyer, who had been in charge of the call-blasting, and Mike Street, who was armed with a Sony camcorder with Night Shot capability, made their way toward the team's location hoping to video whatever had thrown the branch. Shortly before Colyer and Street reached the firefighters, the pair reported hearing the sound of something running away quickly through the brush. The men relayed to Colyer and Street that it "sounded exactly like a man running through the dense ground cover (the scenario was recreated on Sunday afternoon at the site of the branch-throwing incident by the team members)."[40] Due to the odd goings on at the location, two trail cameras were posted at the site.

On Saturday, the teams decided to essentially repeat the activities of the night before, but to shift the "Primate Web" location about three fourths of a mile to the southeast, the direction from which the teams had heard the reply howls the night before.

The evening was a quiet one and team members heard little in the way of known wildlife. Two teams reported possibly hearing Ohio Howl-type vocalizations late in the night, but they were very distant and almost immediately covered up by the sounds of coyotes. Only one team experienced much in the way of unusual activity that night. The team was made up of veteran TBRC investigator Mike Hall and *Palestine Herald-Press* journalist Cindy Parker who intended to write an article about her experiences with the group. This team reported being circled by an animal, possibly two, for hours. The pair was unable to see what type of animal was circling them despite the sounds of footsteps being "very close." The investigator and the reporter endured this stalking behavior for quite a while before they began to hear the animal make a strange and loud call of some kind. Hall said it sounded like nothing he had ever heard in the woods. The best description he could come up with was that the call "sounded like a bleat from a 500 pound fawn." Unnerved, the team requested to be escorted out of the area.

After being notified of the activity, Daryl Colyer, Alton Higgins, and one other investigator began making their way to the team's location. Hall and the reporter had decided not to wait for extraction and had made their way out of the woods to the forest service road by the time the group leadership arrived. The team decided to head back to base camp while Colyer and Higgins opted to go investigate the site where the pair had been circled and had heard the odd vocalizations. The men saw nothing but did hear the odd, bleating call the team had reported. The sound was quite loud at first but faded over the next several minutes until it was heard no more; the animal responsible for the weird vocalization had apparently left the area. The pair heard the call so well but never heard any sound of movement in the dense forest, as the animal seemingly moved away. The men eventually decided that the odd sounds had most likely been made by a fox. Foxes are known to make a wide variety of loud and weird calls/vocalizations; in addition, they are small and light and would be able to retreat through heavy vegetation without being heard. Colyer and Higgins decided to deploy two camera traps and a pair of pheromone chips near the site for the rest of the night in case the animal, whatever it was, returned. They returned to base camp at approximately 3:00 a.m.

These events would turn out to be just a prelude to the events of Sunday, January 9. Team members all rose early and went about the business of packing up their gear, collecting cameras and pheromone chips left in the field, and taking one last look around for potential evidence. Packing up had barely begun when Hall, who had been at the heart of the odd occurrences the night before, dropped a bombshell on operation leader Daryl Colyer.

Hall reported that on the drive back to base camp the night before, he and Parker had encountered a large, bipedal, hair-covered animal walking in front of them on the forest service road. The creature veered to the south side of the road as they drew near and entered the forest. Hall stopped, exited the vehicle, and tried to locate the creature with a flashlight. He was able to spot the animal as it stood partially obscured by a small tree. Parker joined the investigator and the pair watched the creature as it retreated deeper into the woods.

According to the after-action report of Operation Primate Web II that was featured in *The Town Talk* newspaper of Alexandria, Louisiana:

"Parker was the first to spot the creature. She described it as covered with longish black hair and estimated that it stood approximately six feet tall. It had an exceptionally broad body, estimated at about four feet in width. The observation that possibly impressed Parker the most was the smoothness of the creature's movement—'fluid…like it was floating.'[41]

"Interviewed separately, Hall estimated that the creature stood about five to five-and-a-half feet tall. The body was covered with hair ('almost shaggy') described as being very dark brown to black in color. Like Parker, the investigator was also impressed with the creature's thick, powerful-looking body.[42]

"Colyer and Higgins questioned and tested Hall and Parker on every detail in an attempt to rule out any possibility of misidentification. Both witnesses were adamant regarding the possibility that they had mistaken a deer, hog, coyote, dog, large cat, bear, person, or anything else, for what appeared to be a hair-covered primate. Parker, a reporter and Bigfoot skeptic, insisted what they had observed was no known species. Parker was quite emotional and still somewhat in a state of disbelief while she was being questioned. When asked why the two failed to return to the main team (the pair had chosen to pass the rest of the night in town rather than try to sleep in the woods), the investigator and Parker simply replied that they were so shaken by what they had seen that the thought didn't occur to them to return. Hall could not believe that after all that he and the rest of the team had experienced over the last several days, that 'it was that simple' to just drive down the road at 2:00 a.m. and have an encounter.

"Hall attempted to provide directions to the sighting location and described landmarks noted the night before. Finally, expedition leader Daryl Colyer asked if he would simply lead a group to the spot. The investigator agreed and most of the remaining team members accompanied him and Parker to the location. The team found impressions in the pine needles covering the ground in a thick layer along the side of the road indicating the passage of a person or large animal into the dense woods.

"The TBRC tracker at that time followed the trackway for several hundred feet. Tracks were marked with numbered cards. The toe to toe (or heel to heel) step intervals were not particularly impressive in terms of lengths recorded by other researchers. For example, the first set of tracks going up a fairly steep four foot embankment along the side of the road were about 33 inches apart. On level ground within the woods, the step interval averaged about 44 inches in length. Because of the thick layer of pine needles, no clear determinations could be made regarding foot proportions, but the tracks appeared to be about 13 inches in length.

"At one point the creature's path took it next to a pine tree with two dead branches protruding at five to six feet high. The branches were broken off. One of the branches was observed lying on top of a delicate web of vines. The color of the broken ends and the position of the limb gave evidence of the recent timing of the break. The two ends fit together perfectly.

"Fifty feet or so past the tree with the broken limbs, the creature stepped on a small rotting log lying on the ground. The water-soaked log had been crushed by the weight. Team members unsuccessfully attempted to register significant impressions on the log. Only the combined weight of Colyer and another team member (over 400 pounds together) applied in a single step on the log—the team member on the back of Colyer—produced a comparable impression. Of course, it was only as wide as his boot, less than five inches. The crush mark left by the unknown foot appeared to be something in the range of six or seven inches in width. This suggests that it would also have been heavier than the two men's combined weight because of the broader foot and greater surface area."[43]

The team members were simply flabbergasted by the report of Hall and Parker. The whole incident was both incredibly exciting and terribly frustrating. Some members of the group were upset that the pair had not immediately reported the sighting to other investigators. Some felt that had they had been informed, they might have had the opportunity to get on the trail of the animal and capture the video footage the group coveted. Other members felt like the pair's reaction, while frustrating, was understandable. The combination of a shocking–even disturbing–visual on top of nearly 48 consecutive sleepless hours might cause anyone to react in an unexpected manner. Either way, the opportunity to pursue the target species was lost that night.

Just as things were really getting interesting, everyone had to leave. For her part, journalist Cindy Parker did go on to write her article for the *Palestine Daily Herald*. In the piece, she detailed her experiences and concluded: "After my experiences in the field, I have reasons to believe than an unknown creature does reside in the Sam Houston National Forest."[44]

Although Operation Primate Web II failed to acquire any video footage or other evidence, there were some positives to the weekend. First, the group learned that it could effectively pull off a large-scale operation involving many participants. Also, the odd experiences of multiple investigators seemed to validate the practice of call-blasting. The broadcasting seemed to lure the target species almost into the lap of the group as the sighting by Hall and Parker seemed to show. Finally, the group members believed they had proven, at least to themselves, that apes did reside in the SHNF of east Texas and that it was possible to get a look at them. Maybe solid evidence could be obtained in the future.

4

Operation Thicket Probe

Emboldened by the success of several small weekend operations in the winter of 2005, the TBRC leadership began making plans for bigger, more extensive efforts. The next operation would take place over the long Labor Day weekend in the legendary Big Thicket National Preserve of southeast Texas.

In 1974, the United States Congress passed legislation that created the Big Thicket National Preserve. Today, the National Park Service manages over 113,000 acres of the public lands and water resources that comprise the Big Thicket.[45] Incredibly diverse, the Thicket has been called an "American ark" and "the biological crossroads of North America" as it has one of the most biologically diverse assemblages of species in the world.[46] It is estimated that the region is home to approximately 1,320 species of trees, shrubs, vines, and grasses, 60 mammal species, 86 reptile and amphibian species, 34 freshwater mussel species, 1,800 invertebrate species, 97 fish species, and at least 300 bird species.[47] Alligators, white-tailed deer, roadrunners, feral hogs, and coyotes all inhabit the area. Historically, the Big Thicket has been home to bison, jaguars, and red wolves.[48] Animals not recognized by science are also rumored to prowl the Big Thicket. Among them, black panthers and the creature being sought by the men and women of the TBRC, the wood ape.

Reports of the legendary wild man, sometimes referred to by locals as the "raggedy man" or "ol' mossyback," have persisted for decades and date back to the Thicket's earliest settlers. Though most in the region today would scoff at the notion of a monster-ape living in the woods and bottomlands of the area, families who have been living in the Thicket for generations and inhabit some of the more remote areas in and adjacent to it, do not scoff and will, in private, share that they have little doubt that such a creature exists and is present in the area. Those who have managed to catch a fleeting glimpse of one of these animals describe it as resembling a hairy, upright, bipedal, foul-smelling, man-like ape. It is extremely tall, often exceeding seven feet in height. Its loud, raspy, howl-like scream, which is unique and very different from the calls of known animals in the region, has been attributed to a wild man through the years.[49] The group decided to initiate a field study in the Big Thicket after studying historic and contemporary sightings, doing terrain analyses, and discussions with local residents and National Park Service representatives.

Operation Thicket Probe involved less than a dozen members, and was joined by Rob Riggs, a Texas newspaper man, editor, historian, and author. Riggs had a long history researching many of the odd goings-on in the Big Thicket and consented to act as a guide. Also, a film crew from the *Weird Travels* series at the Travel Channel network was scheduled to join the group.

The weather conditions in September in the Big Thicket were quite different than they had been in Sam Houston National Forest at the beginning of the year. Temperatures hovered around 90°F during the day. Nighttime temperatures dropped to around 72°F. The humidity was very high and the mosquitoes were fierce.

At roughly 2:00 p.m. on Friday, September 2nd, the first wave of investigators arrived: Daryl Colyer, Alton Higgins, and Mike Street, along with his Red Heeler canine named Speck, were a seasoned bunch and veterans of multiple wood ape-related operations in the past. The group set up camp at the end of a remote forest road that penetrated roughly five miles into the Thicket. Higgins discovered he had inadvertently left his tent poles at home; he would have to bunk with Colyer in his tent. It was the first of many problems the group would suffer during the operation.

After setting up camp, the three men attempted to contact Rob Riggs and the other investigators but were unable to reach them. Rather than just sit around, they decided to explore the area a bit by walking a pipeline right-of-way that cut through the forest about a half mile east of camp. These pipelines, or *senderos*, were the only cleared areas in this particular area of the Big Thicket. There were no man-made trails of any kind to traverse and some serious brush-busting had to be done in order to even reach the pipeline. The vegetation was so thick, it took the three men nearly 30 minutes to reach the right-of-way despite it being only about 800 yards from camp.

The trio then hiked roughly four miles to the south along the pipeline before heading back to camp. On their return, the men decided to cut through the thick forest to look for wildlife signs, which they found in abundance. Hog spoor, in particular, seemed to be everywhere. The team also walked up on a group of six young pigs during their trek and had a brief standoff with an ambitious young boar. Tired and drenched in sweat, the three men arrived back at base camp at approximately 6:00 p.m. where they found investigators Mike Hall and Josh Daniels*, along with Mr. Riggs, waiting for them.

The group decided to take a passive role that first night and simply stayed in camp. Nothing of an unusual nature was seen or heard all night.

When the Travel Channel film crew arrived on Saturday, the interviews began and continued again after meeting with the National Park Service ranger in a nearby town at 2:00 p.m. The interviews were lengthy and took several hours. Because of the circumstances, important equipment checks went undone, and no preliminary scouting operation was undertaken. These neglected tasks would haunt the operation in the hours to come.

Once the Travel Channel crew wrapped up the interviews at 6:15 p.m., and investigators Jerry Hestand and Monica Rawlins arrived to help with the night's activities, Colyer gathered the group, and he and Higgins laid out the plan. The team would implement LORD (listen, observe, record, and document) protocols on the pipeline right-of-way that cut through an area of interest.

"Riggs was convinced, based on his experiences and reports that he had investigated, that Sasquatches used these right-of-ways as travel corridors," explained Higgins. "We hoped our people would see the animal approaching the call-blasting site by way of one of these clear-cut pipeline right-of-ways."

The group, including the television film crew, took about an hour to make the two-mile hike to the preferred spot on the pipeline. Once there, the members were divided into three teams: Alpha (Colyer, Rawlins, Higgins, and Riggs), Bravo (Street, Daniels, and Hall), and Charlie (Hestand by himself in a tree stand). The television crew was divided up among the two larger teams.

Higgins escorted Hestand to a good spot about 200 yards down the pipeline where he found a suitable tree for his climbing stand; a sturdy loblolly pine. Once Hestand was deployed, Higgins returned to Alpha Team, who would be doing the call-blasting that night. Prior to nightfall, the members of Alpha Team attempted to get a mini-disc audio recorder up and running but had no success. Higgins recalled, "I messed with the recorder for over 30 minutes but finally had to give up. A component was missing, and it wasn't going to work." The group thought that luck was on their side when the television crew's sound technician provided a replacement part, but they then discovered the microphone was bad.

Still, the group had a second recorder with Bravo Team, keeping hope alive that any sounds attributable to the target species might still be recorded successfully "As fate would have it," Higgins recalls, "we found out later that something went wrong with that recording effort as well, and nothing was recorded that night."

A bit aggravated, but undaunted, the team continued preparations to call-blast. Shortly before 11:00 p.m., one of the television crew members noticed some strange white lights down the pipeline right-of-way. Riggs had shared the story of the "ghost lights" of Bragg Road (a dirt road that cuts through the woods outside the small town of Saratoga) during his interview earlier in the day, so many in the group got quite excited by the appearance of these anomalous lights in the middle of nowhere. It was not long before the roars of ATV engines were heard, however, and it became obvious that these were no ghost lights; rather, it appeared to be poachers using spotlights to hunt deer.

Operation leader Daryl Colyer immediately ordered everyone to grab as much equipment as they could and duck into the woods. "I consider poachers to be very dangerous," said Colyer. As the riders got closer, a gunshot rang out, so the group kept a low profile.

The ATV riders stopped right where the group had been sitting minutes before and within a dozen yards of where they were now hunkered down in the woods, but the riders never saw them and moved on.

The poachers then headed toward Jerry Hestand's tree stand. At one point, Hestand decided to video the ATVs. Unfortunately, he was not ready for what happened next.

"He's aiming at us!" one of the ATV riders shouted. It seemed the poachers had mistaken the red LED on Hestand's Night Shot video recorder for the laser site of a rifle.

"Oh, shit! They're going to shoot me," recalled Hestand. "The next thing I know they are pulling up to my perch and stopping."

Hestand quickly realized the group of riders–two to a vehicle–were teenagers. He took an authoritarian tone with them and told them they were on National Park Service land and they needed to leave. Which they eventually did.

Everyone was greatly relieved that the riders were gone, but relief gave way to gloom as the team members realized that any wildlife in the area had likely fled in response to all the commotion. Spirits were lifted a few minutes later when the group began hearing the calls of barred owls. Maybe wildlife was still in the area after all. The group decided to sit quietly for an hour and let things die down before starting call-blasting.

The first few call-blasts went unanswered. The forest was totally silent save for the chirps of crickets. "It is unusual not to hear anything," Colyer explained. "If nothing else, we usually hear coyotes; coyotes usually go nuts when they hear the Ohio Howl."

At 1:00 a.m. after blasting the Ohio Howl again, Higgins and Colyer heard what sounded like a distant reply to the northwest. The reply was identical in pitch and duration to what they had just broadcast. "It was so indistinguishable from what had just been played that we thought it was the television crew's sound technician playing back a recording of what we had just blasted," said Higgins. This was not the case, however, as the technician confirmed that she had only been recording and had not played anything. Colyer quickly contacted Bravo Team via radio. They confirmed that they also had heard the return howl.

The reply vocalization was seven to eight seconds in duration—much longer than what coyotes or owls typically produce—and sounded identical in every way to the Ohio Howl that had been broadcast. The investigators then realized that the television crew's professional detachment had evaporated. They were now every bit as excited as the wood ape hunters.

The back and forth between the call-blasting TBRC members and the howler deep in the forest continued over the next hour. Colyer lowered the volume of the broadcasts intermittently as the mysterious howler got closer. It appeared that the animal was approaching the call-blasting site from the direction of the bayou. Plans were made to intercept the creature and get it on film. But the howler suddenly went silent. More than two hours passed as the investigators scanned the woods and right-of-way with night vision and a thermal imaging unit but they saw nothing unusual.

The team was understandably excited by the night's activities, but lamented the fact that the two mini-disc recorders had malfunctioned. Matters only became more depressing when the Travel Channel crew informed them that, though they had heard the reply vocalizations with their own ears, the calls had still been too faint to be successfully recorded by their equipment. Begrudgingly, the group decided to wrap up operations for the night after retrieving Hestand from his tree stand.

Back at camp, Higgins checked a camera trap he had posted before leaving for the pipeline. He wanted to see if anyone or anything had visited in their absence. The camera had not been activated, but Higgins did notice an odd dusty mark on the television crew's rented Ford Expedition. Higgins had made a practice of checking vehicles for handprints since being part of a BFRO team that had discovered a putative Sasquatch handprint on the roof of an old junk car on Native American land in Concho, Oklahoma, five years before. It seemed his diligence had paid off again as the dusty impression resembled a huge handprint. As Higgins inspected the dusty mark more closely, he saw unmistakable lines in the print indicating the presence of dermal ridges.

Higgins was intrigued but had some doubts about what he was seeing. "There were some weird things about the largest of the dusty marks; even though it appeared to have been produced by a hand, it didn't really look like the possible Sasquatch handprints I'd seen elsewhere," he said. "Those prints were extremely broad, whereas this one appeared to be rather narrow; however, the print was at the top portion of the window, which was quite high on this model of SUV, and it looked as if it may have continued onto the metal of the roof above the window, an area I could not observe clearly. The other strange thing was that this print appeared to be a 'dust on dust' print, unlike what we've seen with other prints where the dust coating on the vehicle is removed after coming in contact with a hand."

The entire team was re-energized by the discovery of the possible handprint. The Travel Channel crew agreed to delay their departure until the odd mark could be examined by an expert; they wanted to film whatever expert the team called in as he examined the possible handprint.

There really was only one person the team even considered calling: Jimmy Chilcutt. Chilcutt was perhaps the world's foremost authority on primate fingerprints at the time. A long-time investigator for the Conroe Police Department, Chilcutt specialized in fingerprint, handprint, and footprint evidence, and lived only about an hour away from the campsite. Once contacted, Chilcutt agreed to make the trip to the Big Thicket to examine the possible handprint.

Chilcutt arrived at base camp about midday on Sunday and was initially quite intrigued by the print. It did not take him long, however, to figure out this print was actually an amalgamation of four human prints involving the side of the palm (the "fat fingers") and a forearm. Chilcutt said, "The 'dust on dust' characteristic of the print is a product of the oils and other skin secretions deposited on the glass by a hand and an arm. As the Ford Explorer was driven down the dusty road, the oils attracted dust to a greater degree than the surrounding glass surfaces, causing a greater build-up."[50] The unusually high position of the faux print near the top of the window of the tall SUV still puzzled the researchers, but Chilcutt felt that was easily explainable as well. He posited that the window had likely not been all the way in the up position when the impression was made, making it more accessible. If so, Chilcutt said, once the window is partially rolled down, and the top part of the window not accessible in the up position is exposed, more of the print should be visible. This proved to be exactly the case. The print, while bearing a striking resemblance to a huge hand, was not evidence of the target species after all.

While the investigators were disappointed, they still insisted the Travel Channel crew film the investigation and Chilcutt's full explanation. The members felt it was important to demonstrate that the group was committed to objective expert analysis of evidence, whatever the results.

The television crew could not stay another night, and Chilcutt and several of the researchers also had to leave due to familial or work obligations. Soon, only Colyer, Higgins, Rawlins, Hestand, Street and his dog remained. While undermanned, the spirits of the group were raised when they somehow got the mini-disc audio recorders working. The group would be ready for any further howler vocalizations.

The team studied a topographic map of the area in order to get a better feel for the region from whence the howls of the previous night had come. All agreed that the vocalizations had come from the far side of a bayou that was just west of camp. The area appeared quite large on the map and was inaccessible by road. The team nicknamed the area "the bubble" because of its general shape on the topographic map. This bubble would be the target of that night's call-blasting operation. First, however, the team decided get some much-needed sleep as they had only slept about four hours in the last two days. But, as it turned out, the group would only be afforded about an hour of rest.

Higgins recalled being awakened from a deep sleep by Colyer at about 10:40 p.m. "Mike Street had awakened Daryl and said that he had heard a loud scream from only 100 yards or so east of our camp," remembers Higgins. "It turned out that Jerry Hestand heard it as well." Higgins rushed out of his tent in order to get one of the mini-disc recorders going. As he wrestled with the audio recorder, he could not help but recall other opportunities to capture possible Sasquatch vocalizations over the last few years that had been missed due to equipment failures or human error. Suddenly, a long, loud scream, starting at a high pitch and descending, erupted from an uncomfortably close distance out in the dark woods.

The entire camp came to life after this second scream. All agreed that the vocalization sounded almost like a female screaming. "It was sort of an 'Aaaaaahhhh' sound; powerful and prolonged and nothing like the 'whoo-ahh' part of a barred owl vocalization and nothing like a coyote call. I got the recorder started," said Higgins. Mike Street asked Monica Rawlins, a fellow investigator and the only female in camp, to scream a reply. She did so, but her screams were a poor imitation of whatever was out in the Thicket. "Her cries sounded feeble and thin in comparison to what all of us had heard," said Colyer. There was no vocal response to her shrieks. Whatever was out there had gone silent.

The group decided to break protocol and call-blast from base camp. The reasoning was simple: whatever was out there was well aware of their presence and had to still be close. Tromping through the woods to get to the bubble now seemed unnecessary when the target of their search had already come to them. Gibbon whoops were selected for broadcast in the hopes that they would incite a reply.

Two more screams were heard, but it was unclear if they were in direct reply to the broadcasts. Team members estimated the creature had moved away from camp and was now 200-300 yards distant. Street, accompanied by Speck, decided to walk slowly through the woods in the direction of the screamer, hoping to get something recorded on his Night Shot-equipped video camera. Colyer played one more gibbon whoop at 2:10 a.m., but there was no response. Colyer joined Street, and the two methodically probed the surrounding woods using night vision. They found nothing.

Once again, something had gone wrong with the mini-disc audio recorder, and none of the amazing screams had been recorded. Colyer remembered the frustration he felt: "I might very well have beaten that recorder into a million pieces if it hadn't been for Jerry." Jerry Hestand had decided to run his video camera during the screaming event, and while it had been far too dark to capture anything on video, he had managed to capture audio of one of the screams. While it did not sound as loud on tape as it had in person, the group was thrilled to have finally captured some audio evidence.

At 2:40 a.m. the exhausted team members began retiring to their tents. Before he turned in, Higgins—ever the optimist—installed a blank disc in the recorder and turned it on. But there would be no sleep for the team, as one of the most intense events in the history of the group was about to take place.

At some point during the night, Colyer woke Higgins and asked, "Do you smell anything?"

Higgins could not detect anything unusual.

A dumbfounded Colyer could not believe his tent-mate could not smell anything. "It was a strong smell, like a damn zoo," he said.

A short while later, something woke Higgins. "I'm not sure what it was," he said. "For a while, I don't know how long, I just lay there in silence."

Then all hell broke loose, as the silence was shattered by what all members agreed was an "astonishing noise."

"It was a savage, huge-sounding, growl/roar that erupted from the forest south and east of Mike Street's tent, and there was no mistaking the proximity of the racket," said Higgins. "It was close and unlike anything I'd ever heard."

Higgins lay there on the floor of the tent, stunned. He tried to make sense of what he was hearing. Was it two animals? The sound was ongoing and constant but did seem to have a rhythm or regularity to it.

"After what seemed like a long time, though it may have only been 15 seconds or so, I realized that Daryl was sleeping through the cacophony," Higgins said. "I hit him with my elbow to wake him and said, 'Daryl, something's got Speck!'"

Daryl Colyer awoke and heard the bizarre and frightening roar-like growl. He heard what he assumed was Speck growling while another, much louder, unidentified animal growled ferociously and continuously in reply. "I think Speck busted him," said Colyer later. "I think that sucker crept up on camp to have a look at us and Speck caught him."

The pair sat up and turned on a bright flashlight and shined it through the mosquito netting of the tent's sloped roof, hoping to see what sort of beast could make such an intimidating sound, but they could see only vegetation. "We yelled repeatedly for Mike," remembers Colyer. "He was only a few feet away from us, but we got no response." From inside his truck, Jerry Hestand was hearing the same terrifying sounds as Colyer and Higgins but had no way of knowing if anyone else was awake. He stayed in place.

Shaken, Colyer, and Higgins continued to scan the woods with their flashlight. Higgins realized the guttural noise likely came from only one creature. He said, "It was as if an extremely large animal was producing sound as it both exhaled and inhaled, because the sound was continuous. This is what caused me to initially believe there was more than one of them." Higgins added, "The sound varied in quality, alternately changing from a very low-pitched rumbling, roaring growl to a slightly higher-pitched sound incorporating sort of a gasping wheeze."

The intimidating noises stopped as suddenly as they had started, leaving the men pondering what to do next. Amazingly, Street and Rawlins had remained in a deep sleep through the entire incident. After the adrenaline left them, Colyer and Higgins returned to their tent, and Hestand to his truck. They were soon overcome by fatigue and went back to sleep. Higgins remembers that his final thought before drifting off was, "I hope that mini-disc recorder is working."

Higgins was awakened one final time a short while later by an animal thumping heavily against his body as it settled on the ground just on the other side of the tent wall. Initially alarmed, he quickly realized that the animal had to be the dog, Speck. She had apparently found her way out of Mike Street's tent and could not get back inside. Higgins theorized she was scared and sought out the reassuring presence of physical contact as best she could.

Once awake, the group rushed to check on the mini-disc recorder. This time, the recorder had functioned perfectly and recorded two full hours of night sounds. Unfortunately, the disc had reached its capacity and stopped recording prior to the arrival of the unseen visitor. The ferocious-sounding growling vocalizations had not been captured. "This series of events is what led us to joke about a Bigfoot curse," Colyer said. "Now, every time some piece of equipment malfunctions, we say the Bigfoot curse got it."

The team dutifully conducted a full reconnaissance of the area east of camp but found nothing that they could positively attribute to wood ape activity. There was nothing more that could be done. It was time to break camp and head home.

Once again, success had been tantalizingly close only to slip away due to equipment limitations, malfunctions, and human error. Still, the call-blasting had proven, again, to be a viable way of getting the attention of, and drawing in, the target species. The team members felt that with better equipment, it was only a matter of time before they obtained the evidence they sought. Also, the group felt that the professionalism and level-headedness exhibited by members would cast the organization in a good light once the *Weird Travels* program was aired. The hope was that the program would encourage other high-quality individuals to join the group's ranks.

In the aftermath of the astonishing events of Operation Thicket Probe, and the failure to record the compelling audio heard during the outing, the group began researching high-end recording systems suitable for outdoor use. Following the advice of experts from the Cornell University's Lab of Ornithology, the TBRC purchased new state-of-the-art components that included a Marantz PMD670 digital recorder, a Hitachi 6GB microdrive to enable uninterrupted audio recording for up to eight hours, a Sennheiser MKE 102S/K6 omnidirectional microphone, and other items to complete a package that would be capable of recording anything audible to the human ear.

5
Operation Thicket Probe II

Excitement was running high after the incredible events experienced by the group in September of 2005 during Operation Thicket Probe. While no video or photographic evidence had been secured, investigator Jerry Hestand had managed to capture one vocalization on his video recorder. This recording was analyzed by an expert in wildlife acoustics, who asked to remain anonymous, but the expert could not identify the vocalization as belonging to any known indigenous species.[51] Though not recorded, the roaring, wheezing, and growling vocalizations heard at close range were strangely similar to the vocalizations of black howler monkeys.

Group members had scarcely been home a day before planning commenced for a return trip to the Big Thicket. But Operation Thicket Probe II would struggle to get off the ground. Access restrictions and safety concerns associated with deer hunting season were a problem. The National Park Service (NPS) decided not to issue the necessary Scientific Research Permit until hunting season was over. More delays would follow, the biggest due to the devastation caused by Hurricane Rita. After much wrangling with the NPS, the necessary permits were acquired, and Operation Thicket Probe II was scheduled to kick off in early January of 2006.

In order to better assess tree damage and high-water levels in the Big Thicket in the aftermath of Hurricane Rita, investigators Chris Buntenbah, Daryl Colyer, and Walter Blake* conducted an aerial reconnaissance flight over the region in December of 2005. The group flew over each of the Big Thicket's units and major watercourses. All the creeks, rivers, and bayous in the region appeared to be at high levels. Little Pine Island Bayou, Turkey Creek, and Village Creek were actually out of their banks in spots. It seemed that combing the Big Thicket on foot, a difficult proposition under normal circumstances, would be virtually impossible in many locations due to the combination of windfall damage and high-water levels. Despite these difficult conditions, the group decided that Operation Thicket Probe II would go on as scheduled.

Since they were operating under a Scientific Research Permit, members were allowed to headquarter out of the Big Thicket Research Station, located in the small town of Saratoga, Texas. The Research Station was equipped with two dormitories—each with ten beds—a full kitchen, restroom and shower facilities, a dining area, a laundry room, a laboratory, and a large conference room that featured topographic, aerial, and satellite maps of the region. To members accustomed to sleeping in tents during field operations, the facility seemed absolutely luxurious.

The goals of Thicket Probe II were the same as those of previous operations, and the equipment was much the same as well. The equipment included night vision monoculars, Sony Night Shot video recorders, thermal imaging devices, mini-disc audio recorders, bionic ears, portable sound-blasting system, game cameras, and Garmin Rino GPS/radio units. In addition, the team would be utilizing the new Marantz digital recorder, a high capacity Microdrive, and the Sennheiser omnidirectional microphone in the hopes of avoiding the issues that had plagued Operation Thicket Probe.

On Wednesday, January 4, once all the investigators had arrived, 18 men and women in total, a briefing was held in the conference room. The plan for this first night was dubbed "The Neches Experiment." Investigator Walter Blake had brought a large, quiet, flat-bottomed boat to conduct an exercise on and along the Neches River, where many sightings of Bigfoot had been reported in the past. Investigators not on the boat would split up and take separate vehicles to drive predetermined roads that ran through the Big Thicket and near the Neches River. Each vehicle would have a mounted video camera running during the exercise.

There were concerns that running the Neches at night so soon after a major hurricane might be a risky proposition. The river was still full of downed timber and other debris as a result of the damage done by the storm. In addition, the water level of the Neches in this area was subject to rapid fluctuations depending upon whether or not electricity was being generated by the Town Lake Dam, known locally as "Dam B." When electricity demand goes down, the release of water from the dam is slowed to a mere trickle, dropping the water level in the river and making navigation hazardous. Despite the possible dangers, the group decided "The Neches Experiment" was a go.

The plan was fairly basic. The boat would enter the Neches River near Evadale. Blake would run the boat north and drop off teams of two to three people on the banks of the river at intervals of a mile or so. Once teams had been deployed, Blake would reverse course and head south. Daryl Colyer would broadcast a combination of known primate vocalizations and the Ohio Howl from the bow of the boat. The teams deployed on the river bank would listen carefully for any response vocalizations and remain vigilant for any other types of activity.

As is often the case in such matters, unexpected difficulties were encountered. Upon arriving at the Evadale boat ramp, the team found the water was much lower than expected. Blake had to back the craft beyond the point where the ramp concrete ended in order to get the boat to float off the trailer. In doing so his truck became stuck in river muck at the base of the ramp and it took more than an hour to pull the truck and trailer out of the deep Neches River mire.

The investigators were finally able to shove off at approximately 11:40 p.m. The plan had been to use as little light as possible (the minimum bow and transom lights required by law) in order to navigate; however, the river was so full of debris and the water level so low that this proved not only impractical but also dangerous. In addition, a misty fog was slowly settling over the river, making navigation even more treacherous. White spotlights had to be utilized in order to traverse the river safely, albeit at a snail's pace. Despite Blake's best efforts, the boat ran up on sandbars multiple times. On one occasion, several investigators had to get wet and push the boat off a sandbar by hand.

By 2:00 a.m. it became painfully obvious that "The Neches Experiment" was doomed to fail. The team finally decided to abort the exercise for the night. The only wildlife interactions that night were a few coyote vocalizations and a handful of beaver slaps. The "roadrunners"—the investigators who had been driving the Big Thicket service roads near the river—had no luck either and nothing to report.

Upon returning to the Research Station in Saratoga, one of the road team members quoted the old *Hee-Haw* line, "If it weren't for bad luck, we'd have no luck at all." I, a new and still fairly inexperienced investigator, entered the room just in time to hear his lament. I was covered from the knees down in river mud and still soaked from the waist down. I noted the road team member's neat and clean appearance and said, "Really?" After a moment of hesitation, the entire room of investigators erupted in laughter. So morale was good and hopes remained high for the rest of the operation.

The day team, which had been scheduled to comb predetermined areas of the Big Thicket beginning Thursday morning, was unable to go out as not enough investigators had yet arrived. Most of the members present had been assigned to nighttime operations. The entire group was called into the Research Station's briefing room at 7:30 p.m. to discuss the plan for that evening.

Group leadership had decided not to push their luck by trying to tackle the Neches River at night again and instead turned to a land-based plan. The Jack Gore/Baygall unit of the Big Thicket, which had a history of sightings, was selected as the target. The group had never conducted research there so the operation was dubbed the "Undiscovered Country Experiment."

Four groups of three members each were deployed at intervals of a quarter mile along the one and only road that went into the unit. One member was deployed alone in a tree stand along this same road. Daryl Colyer and his team would be located in the center of this picket line of investigators and broadcast primate calls via the portable sound-blasting system. Call-blasting began at 11:30 p.m. and went on at intervals averaging 45 minutes in length until the pre-dawn hours of the morning.

But there was little in the way of animal response or activity noted. This was a bit puzzling as the unit had a positively pre-historic look to it and seemed a paradise for wildlife. The only event of any significance occurred when one team reported that they "might have heard some kind of chatter." The team had difficulty explaining what the chatter had sounded like and said it had been extremely brief in nature, very distant, and heard only once. The incident was noted but classified only as interesting and inconclusive.

On Friday, January 6, despite being short on personnel, investigators Chris Buntenbah and Jerry Hestand volunteered to lead a small day team into the field in search of trace evidence. The team combed areas along several bayous and creeks, examined hog wallows, and followed game trails for miles in the boggy, mosquito-infested woods. Creeks, rivers, bayous, and other waterways are natural travel routes for all manner of wildlife. Group members had no reason to believe this would not also be true for wood apes. Despite their best efforts, however, they found nothing of consequence.

The night operations involved call-blasting, but only one team, those to the south of the other researchers, reported hearing any unusual vocalizations. It was described as a "loud and powerful roar" but was heard only briefly before being covered up by a cacophony of coyote yips and howls.

On Saturday, the group decided to head back to an area that had been the source of wood ape accounts in the past, and where they had recorded the howls of an unknown animal in September of 2005. The day teams found nothing of interest in the morning, but the afternoon would prove to be a different story.

In the mid-afternoon, two teams came into contact with some kind of animal that remained hidden in some of the densest vegetation the Thicket had to offer. The researchers could hear the animal move, but their visibility was so hindered by the tangle of vines, hardwoods, pines, briars, logs, and downed trees that they could never see more than a flash of movement. "It was absolutely in the thickest stuff I've ever seen," said one investigator. "I actually got tangled in some vines and fell but never hit the ground. That stuff was so thick it caught me. I was just hanging there, suspended. It was like being caught in a giant spider web."

As the investigators attempted to approach the subject, the animal, whatever it was, only retreated when approached; it never fled. Jerry Hestand failed to get the animal on video. This cat-and-mouse game went on for nearly an hour, but no visual identification was ever made. The team members agreed that, whatever it was, it was quite large based on the sounds of its movements and the apparent ease with which it maneuvered in the thicket. "It could just move in a flash through that stuff," said Hestand. "It took everything I had to move only a step or two."

The night teams focused on the same area, arriving at the pipeline right-of-way at 10:00 p.m. The three separate teams were deployed in a large formation, roughly triangular in shape, with two teams along a half mile stretch of the pipeline. Alpha Team was stationed at, and call-blasted vocalizations of gorillas and gibbons from, the site of the frightening events of the previous fall. Bravo Team monitored the area in which the day team encounter had taken place. Charlie Team including the dog Speck, was stationed where the north-to-south running pipeline was intersected by an east-to-west running *sendero*.

All teams noted some mundane events, such as the sounds of coyotes or the lowing of distant cattle. Then, roughly two hours into the operation, things took a turn.

Sometime between 12:00 and 12:30 a.m., Scott Kessler and Mike Hall of Bravo Team walked over to the edge of the tree line in order to answer the call of nature. While taking care of business, both men saw a large stick fly out of the dense vegetation and whistle past them. Bob Peters*, who was back in the center of the right-of-way, saw and heard the stick land as well. Moments after this incident, the team reported hearing a loud noise as if a large animal or person had slid down a tree. The "slider" landed on the ground with a solid thud or thump. This time the sound of the stick clipping vegetation as it flew through the woods, the sound it made when it landed, the strange sliding sound, and the thud as whatever it was hit the ground, were all successfully captured on one of the mini-disc recorders.

At approximately 1:00 a.m., Alpha Team reported hearing a vocalization very similar to what the Operation Thicket Probe team had recorded back in September of 2005: a call now referred to as the "Big Thicket Howl." The howl seemed to emanate from a position somewhere between their location and the spot where Bravo Team had been deployed.

Investigator Mike Street was informed of the howl by radio and immediately broke off and began cutting through the thick vegetation to meet up with Alpha. If the mysterious howler was in the woods between the now combined Bravo and Charlie Teams and Alpha Team, Street might be able to flush it out.

As the team members were scrambling to get in position, investigator James Morrison* heard an odd sound. "I was the only one to hear it as all the others were walking and whispering to each other." Morrison left his position in an attempt to get closer to the area from whence the odd sound had come. "I heard it," he said. "In succession, a series of about a dozen rapid slaps sounding exactly like a male gorilla (chest drumming). I heard it distinctly." Morrison had his camcorder in hand, but was running it only intermittently in an effort to save the battery and failed to record this very brief event. Morrison turned the camcorder on and continued pursuit. In doing so, he left his position near the crude trailhead and was not in position to capture one of the more significant moments in the group's history that would occur just a few minutes later.

It should also be noted that the reason more of us did not continually run video cameras, phones, and audio recorders has to do with battery life. Trying to recharge equipment in the field was difficult to impossible. Often, individual incidents occurred so quickly that there was not time to power up video equipment before the animal was gone.

The members of Alpha Team then began making their way toward the right-of-way while Bravo and Charlie held their ground. The hope was that Alpha Team would flush out the mystery animal as it attempted to cross the *sendero* within sight of the investigators spread out in that area. Shortly after Alpha Team began the maneuver, Peters and Kessler, now the two investigators closest to the game trail intersection, reported hearing "a very deep, guttural growl" followed by "a very resonant and deep" vocalization. The disturbing sounds were coming from an area just a few yards down the game trail where Bravo and Charlie members had dumped their packs and equipment prior to forming the picket line.

"I immediately focused my attention on the area and observed a very dark, large shape moving within the intersection," said Peters. "I observed it stop at a location that would have put it only a few feet away from the pile of backpacks and gear that we had discarded."

Peters suddenly stiffened. Kessler described what happened next: "As I approached the trailhead, Peters, who was to my left nearest the trailhead, stuck his right arm and hand out across my chest in a 'stop' motion." Kessler then followed the gaze of Peters to the area just down the game trail where the gear had been set. "I did see movement. My best description would be shadow movement: no outline, no features, just the movement of a large shadow."

The men were unsure of just what they should do. They were afraid that any movement or sound they made would provoke the creature or scare it away. "Peters wanted to light it with his white light, but I said, 'No, wait, if you do then we will lose our (natural) night vision and possibly scare it away.'" Peters, a law enforcement officer, and Kessler, a firefighter, held their ground until another intimidating, rumbling growl was directed their way. At that point, the men decided that caution might be the wise choice. "We perceived that it felt cornered and that it would probably be best to back up and give whatever it was some room," said Kessler. As Peters turned slightly in order to step away from the trailhead, the creature bolted.

Kessler recalls: "I saw the skyline darken and go black, just for a few seconds. I thought it was coming after us. I said, 'Oh, shit!' and ducked slightly. It moved left to right; we were facing north so it would have been west to east. I then heard a snap sound to my right at about the two o'clock position. All I saw was a large, fast shadow move across the pipeline right-of-way and into the trees. I do know that whatever it was, was blacked out and was taller than a mile marker in the middle of the pipeline as it crossed." The mile marker would later be measured at just under eight feet tall.

Peters was looking at Kessler when he saw the veteran investigator take a step back, his eyes as big as saucers. "That damn thing just crossed!" yelled Kessler. The other investigators nearby had not seen the animal cross the right-of-way but did hear something burst into the brush and continue to the east. They quickly gathered at the trailhead intersection and began shining red lights into the brush in an attempt to spot the creature, but felt that whatever it had been, it was now long gone as the sound of something tearing through the brush grew fainter. Within moments, Colyer and Street arrived on site and searched the thicket for any sign of the animal, but came up empty.

I had been positioned approximately 75 yards from the spot where the game trail intersected the *sendero* and from whence the creature bolted past

Kessler and Peters. I wish I had seen the tall, upright figure sprint across the pipeline, but I did not. It was too dark, and I was too far away; but I did hear something crashing through the forest to the east at almost the exact instant the headlamps and flashlights of the other investigators came to life. I did see the demeanor of both Kessler and Peters mere moments after this incident took place. Whatever they saw rattled them to their core, there was no doubt about that.

Kessler is a firefighter who is trained to run into burning buildings. Peters, at the time, was a narcotics officer with the Texas Department of Public Safety; the kind of guy who kicks in the door before storming a crack house. These were two tough and rugged men who were not easily frightened, yet both of them were wide-eyed and trembling. Kessler was talking a hundred miles an hour, adrenaline flooding his body. Peters was the exact opposite, almost in a stupor. It took at least a half hour to settle the men down enough to start making our way back to the research station.

Weary, we gathered our equipment and trudged back through the overgrown remnants of the game trail to the vehicles. We were all heartbroken that yet another opportunity to document the subject had been lost. We arrived at the Research Station at roughly 4:00 a.m., cleaned up, and hit the rack. By 2:00 p.m. the next day the Research Station had been cleaned up and emptied. Operation Thicket Probe II was over.

Alton Higgins, now Chairman Emeritus of the NAWAC, was sufficiently impressed with the events that took place during Operation Thicket Probe and Operation Thicket Probe II to write the following commentary on the investigations in May 2006:

"I was present in early September 2005, when many of the events of Operation Thicket Probe took place. Words do not suffice to convey an accurate or adequate sense of events that, in themselves, often serve as compelling evidence for participants, but which for many readers or skeptics merely comprise additional fodder for the fantastic. This is true even when observations are accompanied by support in the form of evidence (which is, admittedly, typically unidentifiable or disputable). Such is the case with the circumstances of Operation Thicket Probe II. In private face-to-face discussions with individuals present on that final Big Thicket night in January 2006, convictions ran near absolute concerning the presence of a huge, fast, bipedal creature, something undocumented by science.

"I've been privileged to have met and spent time in the field with most of the TBRC investigators who experienced the remarkable events described in this report. They are not overly imaginative or inexperienced juveniles; they are solid, sober, mature, courageous, educated individuals from various professional fields who are committed to accuracy and legitimacy. Critics will, no doubt, assert that the events can be explained by means of hoaxing, erroneous perceptions, or outright fabrication. Nobody who, like me, knows the people involved harbors any such misgivings. That is not to say that investigators cannot make mistakes or misinterpret events. Mistakes are inevitable, although 'mistakes' are often identifiable as such only with the benefit of hindsight, as is the case with Operation Thicket Probe II.

"Hoaxing by outsiders, those not affiliated with the TBRC, can be ruled out, in my opinion, for several reasons. For starters, almost nobody outside the TBRC and the National Park Service knew of the group's presence in the Big Thicket Preserve. Participants did not know of the plans and the locations for daily activities until shortly before they took place, making it unlikely that hoaxers could have learned of them, particularly in light of the limited cell phone reception in the area. One way access into the remote locations entered by the research teams would have made it impossible for hoaxers to arrive undetected.

"Since a conspiracy on the part of the nighttime team members to fabricate an elaborate story can be categorically excluded as an explanation, misperceptions and misidentifications remain as the most reasonable alternatives to the proposal that a sasquatch was detected by TBRC team members.

"There is no way, of course, to positively preclude the possibility of erroneous interpretations. The long, tremendously loud howls that have so far defied identification could possibly belong to a known species. The deep, resonant, guttural growl could have come from an unseen predator. The odors detected by investigators in association with unusual activity could be the result of unrelated coincidental events. It is possible that team members could have heard a black bear, or some other large animal, sliding down a tree trunk (although, for the record, black bears were extirpated from this region in the 1920s, and their return has not, as yet, been documented).

"However, it is hard to imagine a conventional explanation for the large stick that came hurtling towards the men from out of the forest (followed moments later by the apparent sound of something sliding down a tree). The act of throwing requires a hand with a thumb; naturally falling sticks would not shoot out across an opening. It should be noted that such behavior has long been reported in association with sasquatch activities. Sasquatches are reputed to throw such things as rocks, pine cones, nuts, sticks, tree branches, and feces, to name a few. The great apes have been documented exhibiting similar behavioral characteristics during intimidation displays.

"Similarly, after reading their accounts and visiting with them, I cannot come up with a feasible alternative explanation for the massive entity seen at close range by these men of high repute and credibility. They saw something. They were close to it. They saw it standing; they saw it moving swiftly. They heard it as it departed. They know it was not a member of their party; everyone was accounted for, and it was far too large. If I strained credulity to its limits, I might propose that they heard (the "howl?") and then saw a barred owl flying fairly low to the ground and across the pipeline right-of-way. This idea would not, of course, explain the sound of a snap heard as it left, or the deep growl that first alerted them to the creature's presence, or the manner in which the creature moved, or the size of it. It seems less incredible to propose that the men heard, saw, and possibly smelled a sasquatch than it is to write off the entire evening's events as the product of a series of improbable and unrelated occurrences.

"One thing worth noting, I believe, is the apparent interest evidenced by the purported wild man in the backpacks. Those familiar with sasquatch reports will recall other such similar accounts. It would appear to be possible that the TBRC investigators were under observation and that when they walked away from their stash of gear, the creature saw an opportunity to examine the backpacks. This could, in turn, imply familiarity with backpacks and their contents. The possible fascination with and momentary distraction provided by the gear and packs may have afforded Kessler and Peters the few seconds of chance visual contact. Such a scenario also seems to be a common thread among sasquatch reports.

"The fact that possible Bigfoot activity has been noted in the same location on two occasions, four months apart, could be interpreted as an indication that at least one individual inhabited the area for the duration of that period of time and that it is possibly a resident. Although we remain convinced that large-scale field operations such as 2005's Operation Thicket Probe and 2006's Operation Thicket Probe II have their place, the fact remains that many hundreds of hours of volunteer activities have produced mere seconds of observations and no compelling photographic images. Accordingly, beginning in April, 2006, the TBRC has been engaged in an on-going long-term project called Operation Forest Vigil, conducted in accordance with the National Park Service guidelines, to focus as intensely as possible on securing photographic evidence from the area."[52]

6
Operation Forest Vigil

For the second time in less than a year, the group had come painfully close to capturing video evidence that might have been convincing enough to pull the Sasquatch out of the realm of fairy tales and into the science books. "Discovery Day," as it was deemed, seemed to be close, possibly as soon as the next group operation. But just as the TBRC appeared poised to make the discovery of the century, the group began to experience a multitude of problems that stunted their documentation efforts.

One issue that is always a factor when dealing with an all-volunteer organization of any kind is getting all members to be equally active and engaged. It seemed that 80% of the work was being done by 20% of the people. There were various reasons for this: family situations, financial problems, job issues, long travel to and from the study areas, the poor physical condition of some members, and the fact that it was practically impossible to schedule an operation on a date when the majority of members would be available. The situation put a strain on the handful of members who were carrying out most of the field investigations and operations, and indeed several planned field operations had to be canceled due to low levels of membership participation.

If that wasn't challenging enough, a change in leadership at the National Park Service (NPS) made things even more difficult. The Superintendent who oversaw the Big Thicket had always been cordial and civil. That is not to say he was on board with the efforts of the TBRC to document a huge ape, but he seemed to recognize that the members of the organization were striving to be citizen scientists in every way and were serious about the subject. He had always insisted that the group secure the proper scientific research permit to operate in the Big Thicket and did not allow the group to conduct outings during deer season for safety reasons, but he never sought to impede the group's overall mission. As is often the case with government personnel, that Superintendent moved on to a new position and was replaced by another gentleman with whom it was a bit more difficult to work. Suddenly, the NPS was slower to act upon requests for scientific research permits, and when they did eventually respond they often wanted group leadership to "correct" or "clarify" something in the proposal and re-submit the application. This made nailing down hard dates for field operations all but impossible.

The group would have to come up with something other than large-scale field operations in order to get the job done while placating the increasingly uncooperative NPS. Also, while short-term operations had yielded tantalizing opportunities and enough anecdotal evidence to convince any and all group members that the Sasquatch was a creature of flesh and blood, they had failed to produce the incontrovertible evidence necessary to prove the existence of the wood ape to mainstream science.

The ideal solution would be a sustained field study where investigators were in the field 24 hours a day, seven days a week. But the group had trouble getting even a dozen people out in the field over a long weekend, nor did it have the financial wherewithal to pay for full-time researchers; even if it had, the NPS would never approve such a project.

It was about this time that the group became aware of the work of Ken Goldberg, a professor of Industrial Engineering and Operations Research at the University of California, Berkeley. In response to a possible sighting of an Ivory-billed woodpecker in the river bottoms of Arkansas in 2004, Goldberg developed an automated birdwatching robot that utilized two video cameras to continuously scan the skies in an effort to capture footage of the bird, a species not seen in the wild since 1944 and long thought extinct.[53] The robot utilized software that scrutinized the captured footage and discarded any shots that did not fit the profile of the woodpecker being sought. The unit was capable of running continuously for months at a time. Once deployed, the system was virtually self-sufficient, eliminating the need for a large team of searchers.

TBRC leadership wondered if Goldberg's robots might be a way around the group's manpower issues. Goldberg was well aware of the problem: "The problem with field biology is that it is very difficult. You have to go out to somewhere remote; it's lonely, it's cold (or hot), it can be downright dangerous, and the presence of a human observer can affect the behavior of the animals you are trying to study. So, our idea is that robots can help."[54] The question was: Would a less intrusive effort along these lines be more palatable to the new Big Thicket Superintendent?

The group was further convinced that some kind of camera or video study would be a viable way to document the species based on the comments of George Schaller, one of the world's leading field biologists, and Duane Schlitter, who at the time served as head of the Texas Parks and Wildlife Department's (TPWD) Non-game and Rare and Endangered Species program.

When questioned about the possible existence of the yeti and/or Bigfoot, Schaller said, "A hard-eyed look is absolutely essential. The best thing to do would be to set up camera traps that automatically take pictures of animals. If this is monitored for a year you may get nothing, but you may end up with some very interesting wildlife pictures."[55]

Then, when questioned about what sort of evidence would be sufficient for the TPWD to officially recognize the Sasquatch as a real animal, Schlitter said, "To conclusively prove…Bigfoot in Texas, we (TPWD) would need an image (photographic or video) that included details to show us that it was not doctored or edited in any way…[56] And so, inspired by the work of Goldberg and the comments of Schaller and Schlitter, planning began in earnest for a new and different kind of operation, one that would be centered around the use of trail cameras.

Operation Forest Vigil would become the most ambitious project the group had ever attempted and involved almost the entire membership, though only a half dozen or so would be needed at any given time. It would encompass three separate study areas, dubbed Areas X, Y, and Z. Area X was located in the Ouachita Mountains of southeast Oklahoma, Area Y was the moniker attached to the particular area of the Big Thicket National Preserve where so many exciting events had taken place, and Area Z was located on private land in east Texas that bordered the Trinity River in Houston County. The addition of two new study areas was exciting; leadership felt that the group could take on the new study areas and do them justice using the new methodology.

After much prep work, Operation Forest Vigil was initiated in Areas X and Z in April of 2006. Alton Higgins, Daryl Colyer, and Chris Buntenbah deployed a half dozen high-quality game cameras, along with a Wildlife Eye video system, on private land in the Ouachitas of southeast Oklahoma where sightings of ape-like creatures had occurred for decades. Other members took the lead in east Texas and deployed four game cameras on a piece of promising land bordering the Trinity River. The plan was to get in, set the cameras up, and get out as quickly as possible so the environment returned to normal. Researchers would return every 90-120 days to retrieve images and refresh batteries. The group was fortunate in that the two new study sites were on privately owned land, requiring only the permission of the landowners to place the cameras. But the main target of Operation Forest Vigil still remained; that required getting permission to place cameras within the Big Thicket National Preserve, which was going to be more difficult.

While the group pondered how to go about pitching Operation Forest Vigil to an increasingly unsympathetic National Park Service, work continued in Areas X and Z. Alton Higgins and Daryl Colyer made the first camera maintenance trip into Area X in July of 2006. What they found was disturbing. All of their cameras, every single one, had been damaged, and some completely destroyed by black bears (the culprits were identified by the photos pulled from the surviving memory cards of the cameras). The group should have anticipated this outcome, as it was known that the Ouachitas hold a substantial black bear population. Unfortunately, the majority of the TBRC membership was from Texas and had no experience with bears; the species had been extirpated from all but the western-most areas of Texas in the early 1900s. It turns out that black bears are attracted to petroleum-based products,[57] and the housings of the game cameras were made of a petroleum-based plastic that proved too appealing to resist for the olfactory senses of the area's black bear population.

The destruction of the first cameras in Area X necessitated the protection of the expensive replacement cameras. All subsequently deployed cameras were encased in heavy metal "bear boxes" that were secured with chains and padlocks. Unfortunately, the metal boxes, chains, and locks were heavy and, though necessary, made deploying the new camera traps much more difficult. The most disappointing aspect of the situation was that, once encased in the metal bear boxes, the small and unobtrusive cameras had a much larger profile and were very noticeable. "We made every effort to camouflage and conceal them," said Daryl Colyer. "We used camouflage-patterned burlap, small limbs, brush, and leaves, but they still stuck out like a sore thumb." Members felt the odds of capturing a photo of the target species had been drastically reduced by the obtrusive bear boxes but hoped for the best and turned their attention back to the Big Thicket.

A proposal was written up and submitted to the National Park Service in February of 2007 requesting permission to place up to a dozen game cameras in the Big Thicket National Preserve. The proposal stressed that the entire operation would be funded privately and all the TBRC really needed was permission to access the study site and place the cameras. The proposal also pointed out that, should the TBRC be successful in its efforts, they hoped to work closely with the NPS and the Texas Parks and Wildlife Department to secure new habitat, conserve existing habitat, and facilitate official scientific and governmental recognition of the North American wood ape. The group proposed a start date of March 1, 2007, for the project.

The group was pleasantly surprised when the NPS quickly gave the project the green light. So in early March, five investigators and I set out to deploy 13 camera traps in the Lance Rosier Unit of the Big Thicket National Preserve. Fully loaded, each of our 75-pound backpacks contained a trail camera, bear box, length of chain, heavy duty padlock, water, and food. We trudged for hours through some of the thickest and most inhospitable terrain I have ever traversed. Thorny vines, wasps, and venomous snakes seemed to be everywhere. Long stretches of the hike involved knee-to-waist-high swamps or bayous. The possibility of stepping on or passing too near an alligator lying on the muddy bottom of the marsh was all too real. Perhaps the most imminent danger was the possibility of stepping in a hole and snapping an ankle. We were lucky no one got hurt, as hauling someone out of there would have been a truly arduous undertaking.

One thing that was encouraging to us was that there was absolutely no sign of human activity once we got about a half mile into the unit. Even the hunters, it seemed, did not venture far into the Big Thicket. We felt this was an encouraging sign as the lack of human activity made this particular unit of the Big Thicket an ideal place for a furtive species to hide.

After much effort, we succeeded in placing the cameras in a semi-circle arrangement off of the pipeline right-of-way where a large animal had crossed in front of investigators during the final night of Operation Thicket Probe II. Hopes were high.

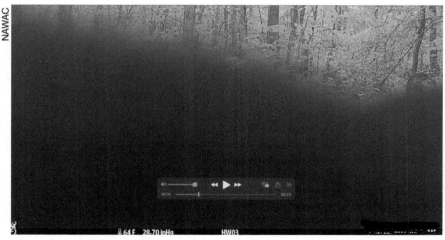

A large, hair covered animal tampering with one of our trail cameras. Such images were common during Operation Forest Vigil even in regions where black bears have long been extirpated.

As Operation Forest Vigil went on, investigators living in central and east Texas took over camera maintenance in Areas Y and Z while members living in north Texas or Oklahoma concentrated on the cameras in Area X. I was living in central Texas at the time but had grown up in southeast Texas. I knew the area well, so I placed myself with the Area Y teams. As time went on, the trips into the Big Thicket became more difficult. Three hurricanes—Rita in September of 2005, Humberto in September of 2007, and Ike in August of 2008[58]—ravaged the area, changing the landscape considerably. Downed trees were everywhere, and an area in which it had always been difficult to hike became almost impossible to traverse.

"It is such a different place now, it's like it is trying to keep us out," I recall saying to fellow investigator Jeremy Wells on one of the post-hurricane camera maintenance trips.

Operation Forest Vigil ran from 2006 through early 2011. At that time, yet another new Big Thicket superintendent refused to renew the group's scientific research permit. Without the permit, the TBRC could not continue to lawfully operate in the Big Thicket. While disappointed, Operation Forest Vigil was winding down anyway. Area Z along the Trinity River had proven to be a bust. Over time, many in the group came to strongly suspect that the reports of activity on the property had been fabricated. Meanwhile, no photos of the target species had been captured in Area X either. The obvious jutting profiles created by the metal bear boxes were believed to be the most likely cause for this as all manner of strange things had been experienced by members during camera maintenance trips over the years; so much so that the group began to feel that Area X might be a better study site even than the Big Thicket.

In all, the TBRC spent approximately $50,000 and countless man hours on Operation Forest Vigil. A few intriguing images were snapped, most of which showed some kind of hair-covered animal so close to the camera that positive identification was impossible, but no definitive photos of wood apes were obtained. I suppose that fact makes Operation Forest Vigil a failure, but the group gained so much from the effort.

We learned the areas inside out where cameras had been deployed. This information was filed away and has proven to be very useful in the years since the project wrapped up. The group also captured thousands of images of indigenous wildlife over the life of the project. In essence, the TBRC provided the National Park Service and the private landowners of Areas X and Z with a free wildlife census.

Finally, in my mind at least, the years of Operation Forest Vigil were a kind of golden age for the organization; a time when the group differentiated itself from others in the field and established a reputation as a hard-working, logic-based, and scientifically-minded group of men and women. Bonds and friendships were formed over these years that remain strong to this day: friendships for which I will be forever grateful.

7

College Professors and *MonsterQuest*

There were times when the TBRC/NAWAC had multiple things going on simultaneously. This chapter describes two other events that took place during Operation Forest Vigil, the large-scale camera trap project that ran from 2006-2011.

In 2006, the TBRC was contacted by three sociologists who were doing research on the types of people who believe in the paranormal and their belief systems. While no one in the group felt that Bigfoot had anything to do with the paranormal, the leadership felt that working with these professors might be a good way to demonstrate to serious academics that we were not a bunch of kooks. Arrangements were made for the three sociologists to participate in a short weekend outing in east Texas with several members of the group in December of 2006. Investigator Daryl Colyer and I met Christopher D. Bader, an Associate Professor of Sociology at Baylor University; F. Carson Mencken, Professor of Sociology at Baylor University; and Joseph O. Baker, Assistant Professor of Sociology at East Tennessee State University.

Colyer, the sociologists, and I met fellow investigators, physician Ken Helmer and Tod Pinkerton, at a hunter's camp in the Sam Houston National Forest just outside of Coldspring. Colyer explained to the professors, decked out in brand new camouflage clothing fresh off the shelves of the Waco Wal-Mart, that we were going to try to get the attention of any apes that might be in the area by call-blasting the Ohio Howl and other vocalizations of known primates.

While call-blasting was nothing new for us, this time we would be using a new device to transmit the audio. In the past, the call-blaster itself had to be directly operated by a person. Even if an ape was attracted by the audio and came closer to the source in order to investigate, it would likely remain concealed once it realized a human was nearby. Helmer had a new remote-controlled call-blaster we would be trying out for the first time.

Around 10:00 p.m., the group drove out to an area through which the

Lone Star Hiking Trail runs and where Helmer had heard wood-knocking in the recent past. The nighttime temperatures were in the high 20s and dropping. We all wore heavy coveralls, wool caps and socks, and gloves, but the professors were not nearly as ready for the weather and shivered noticeably. They never uttered a word of complaint, however.

Our plan was to place the call blasting unit along a remote portion of the Lone Star Hiking Trail and position the speakers in the crook of a tree. The team of investigators would spread out along the trail, away from the blaster, in a sort of picket line in order to listen, observe, and record anything of interest. The hope was that an ape, preoccupied by the sounds being broadcast, might walk right by one of the concealed pairs of investigators as it closed in on the audio's place of origin.

As the professors would later write in their book, *Paranormal America: Ghost Encounters, UFO Sightings, Bigfoot Hunts, and Other Curiosities in Religion and Culture*: "It is an eerie experience to crouch by the side of a path, deep in unfamiliar woods on a bitterly cold evening, waiting for a terrifying howl to erupt from the call-blaster. Between each blast, Keith [the professors used pseudonyms for us in their write-up of the night's events] would wait for responses from the local wildlife and then try again. For those of us who had no idea when the next blast would come, these waits were nerve-wracking."[59]

In a little over an hour of broadcasting calls, the team heard nothing in reply but the bellowing of a distant bull. So the investigators emerged from their hiding places and started walking back to the broadcasting location, not being particularly quiet, when things suddenly got weird.

Professor Baker and I heard something moving in the woods not too far off the trail, something of significant size just out of sight behind the tree line. When we stopped, it stopped. When we started to move, so did the unknown animal. We had played this stop-and-start game for several minutes when we heard what sounded like a man stumbling in the thick brush. After the initial "tripping" sound, there was a quick succession of "crunch, crunch, crunch" sounds as if the animal was trying to quickly regain its balance. I pointed out to the professor that this was not the kind of sound I would expect a quadrupedal animal like a deer or hog to ever make.

While pondering our next move, we were suddenly enveloped by an absolutely terrible smell. It is hard to describe how foul this odor really was. It was similar to the smell of a large outdoor dumpster cooking in the Texas heat after a rain. It was sour, pungent, and quite unlike anything I had smelled in the woods before. I informed Colyer via radio that we had some kind of smelly visitor in close proximity, and once at our location, he and I left the two wide-eyed academics on the trail and plunged into the woods to see if we could locate the source of the smell.

We had not been at it long before we were overcome by an "...obnoxious odor smelling very much like an animal that had been immersed in garbage."[60] We could not locate the source of the odor. What was strange was that there were times while Colyer and I were looking in the woods that we could smell it, but the professors, mere yards away on the trail, could not and vice-versa. It was very odd and seemed to make an impression on our guests, who later wrote that they "...indeed heard what sounded like something knocking on a tree and briefly noticed a foul odor, akin to an animal carcass. We could not locate the origin of the smell, as it came and went."[61]

When it became obvious that whatever had been present was now gone, we packed up our gear and headed back to our frigid camp. In the morning at a diner in Coldspring, over a hot breakfast of scrambled eggs, sausage, grits, and hot coffee, the group discussed the events of the night before. We felt it would be a good idea to revisit the area and look for spoor and tracks in the light of day.

While still quite cold, the morning felt almost tropical when compared to the night before. Professor Baker joined Helmer and myself, forming one group, while Professors Mencken and Bader formed a second group with Colyer (Pinkerton had left earlier that morning). After only about 20 minutes of looking, Colyer contacted our group via radio and asked us to come down to the bank of a small creek that meandered through the area.

Once there, he said to us, "Look at this and tell me what you think." Colyer pointed to the wet, packed sand right on the water's edge. There, clear as day, were two footprints. To the left of the prints was a third impression that resembled knuckles. The tracks were very fresh, no more than a day old at the most. While not impressively large–they were just over 12 inches in length and 5 inches in width–they were unmistakably barefoot tracks. Toes were clearly visible on the print of the right foot. The toes were not visible on the track of the left foot as this print extended slightly into the running water of the creek. The prints lacked any sign of an arch and were impressed a solid inch into the packed sand (the impressions we made while walking the bank quickly disappeared), indicating the creature had been heavy.

Colyer looked up at the professors and said, "It's not a raccoon, a fox, a horse, or a hog either. There are only two things it could be: a human or a Sasquatch."[62] While purely speculative, it certainly was not hard to imagine a scenario where a young ape, perhaps lured to our location by the call-blasting, had stopped, squatted down at the edge of the small creek, supported itself with its left hand, and scooped water to drink with its right. The three impressions were cast with plaster and remain in the NAWAC evidence collection today.

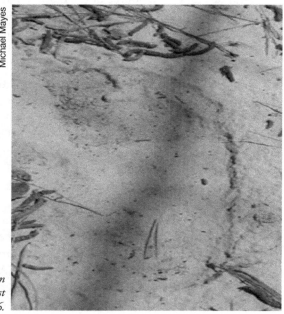

A pristine 14-inch track found in the Sam Houston National Forest in 2006.

In late 2008, the TBRC had another significant media opportunity. Whitewolf Entertainment invited the group to appear on an episode of the popular History Channel series *MonsterQuest*. The episode would center around the legend of the ape-like creature said to roam the swamps and bogs around the small Arkansas town of Fouke. The Fouke Monster was made famous by the cult film *The Legend of Boggy Creek* back in the 1970s, a movie many current researchers cite as the origin of their interest in the Bigfoot phenomenon. Sightings of the beast continue to the present day, and *MonsterQuest* wanted to go looking for it. Up to that point, the group's experiences with television had been positive. The Travel Channel episode filmed in the Big Thicket in September of 2005 had spurred interest in the group and its mission. The positive way in which the TBRC had gone about its business struck a chord with many people and requests for membership had jumped as a result. *Mon-*

sterQuest had proven to be serious about the topics investigated over its first two seasons, and leadership felt that doing the show would be a great opportunity to show a nationwide audience the organization's serious-minded and scientific approach to Bigfoot research.

The "Swamp Stalker" episode first aired on February 18th, 2009, and featured many luminaries of the cryptozoology world: Loren Coleman, Jeffrey Meldrum, and Rick Noll, to name a few. The introduction set a very dramatic tone. Shots of small-town Fouke, Arkansas, are accompanied by the following voiceover: "Fouke, Arkansas. A small sleepy little town with a church, a school, and a local store. Fouke is also home to a nightmare at times so real and so terrifying that some residents packed up and left in the middle of the night. Fouke is the birthplace of the swamp stalker of Boggy Creek."[63]

No one in the group had ever heard the Fouke Monster called the "swamp stalker" before, but everyone figured the producers were running out of Bigfoot-related titles, had run into copyright issues, or both. In any case, investigators Jerry Hestand, Mark Porter, Daryl Colyer, Alton Higgins, and Ken Stewart rendezvoused with the *MonsterQuest* film crew and producers near Fouke. The group was split in two. Colyer, Higgins, and wildlife tracker Mark Peterson were ferried approximately 75-100 miles northwest (though the narrator said the group was 250 miles away!) to an area in the Ouachita Mountains of southeastern Oklahoma, where Alton Higgins had located an impressive trackway some years before. This area would soon become known as Area X.

The second team stayed in the vicinity of Fouke and explored Boggy Creek, the Sulphur River Wildlife Management Area, and Mercer Bayou. The Arkansas unit was composed of Jerry Hestand, Ken Stewart, Mark Porter, and veteran researcher Rick Noll. The episode would go back and forth between these two teams while interjecting a bit about the history of the Fouke Monster and a few contemporary reports of encounters.

The Oklahoma team was shown deploying the most up-to-date trail cameras available at the time. Both Reconyx and Buckeye cameras were used, but the producers zeroed in on the Buckeyes due to their capability of transmitting images and/or video to a remote site up to two miles away. The team could monitor images in real time and react quickly should an interesting photo be captured. But there was frustration between team members, who wanted to put every second to good use, and the film crew who wanted to stage shots, or to have the researchers repeat an activity or a conversation so as to get it all on tape from the right angle. There would be very little opportunity to achieve real results under these circumstances.

Meanwhile, the second team back in Arkansas was experiencing their own frustrations. Jerry Hestand, Mark Porter, and Ken Stewart were avid kayakers and, along with Rick Noll, had devised a plan where they would paddle out along different fingers of the bayou, sit quietly armed with night vision monoculars and video cameras, and observe while Noll call-blasted the Ohio Howl and other primate vocalizations. The idea of observing from the water as opposed to attempting to seclude themselves along the bank was novel.

The producers thought this was a great plan but wanted to add something to it. They wanted to spray a small particle reagent (SPR) on the brush and trees along the banks of the bayou. The SPR would "latch on" to any oils left behind after an animal touched a treated piece of vegetation. An area that had been touched would glow when exposed to ultraviolet light, making it clear that something had been stalking the banks. Hestand, Porter, and Stewart were charged with spraying the SPR on the vegetation along the edges of the bayou and quickly soured on the idea. There were only a few bottles of the reagent and countless plants and trees to treat. The men also questioned whether or not the species could be identified based solely on the glowing chemical. Only after this was completed—a process that took hours—did call-blasting begin.

The episode shifted back to Oklahoma where the most intriguing event to take place was the downing of a trail camera by a black bear. The crew also had Alton Higgins recount how he found a trackway in the area in 2001.

"I saw a little concentration of leaves in this flat, muddy area," Higgins said. "I bent down and picked out the leaves, and a track was revealed." There were multiple prints under the thin cover of leaf litter, revealing a trackway. "They were 16-inch tracks, pretty deeply impressed in the mud. Parallel to the tracks were some bear tracks, an adult and a young bear. So, you could make a side-by-side comparison."[64] The producers had Higgins re-enact his track-find for the camera, something the veteran researcher was not keen on doing. He felt that unless presented very clearly, the impression would be that the tracks had been found while the show was filming.

Back in Arkansas, there had been no discernible wildlife activity in response to the call-blasting the night before. Neither were animal visitors detected after scanning the SPR-treated foliage along the bayou with ultraviolet light. So the show shifted gears and focused on a new type of game camera that took a $360°$ photo when triggered. Rick Noll explained on camera how it worked, but the actual deployment of the device was not shown, and the resulting photos did not effectively demonstrate what the device could do.

While Noll discussed the 360° camera, Hestand, Porter, and Stewart struck out to look for some sign that a curious ape might have approached the call-blasting location the night before. The trio did locate a track of some kind, but it was old, degraded, and left the men unimpressed. The producers, however, wanted to film the team examining the print. As is often the case, the track did not show up clearly on camera, making the investigators appear as if they were examining little more than a roughed-up patch of soil.

In the Ouachitas, the producers resorted to having Colyer and Higgins recount some of their experiences in southeast Texas. Colyer recounted his first sighting of what he believed to be a wood ape along the Trinity River. (Later, actors would recreate the event for the show.) He recalled how he felt the creature might have been attracted to the laughing and squealing of young children who were on the opposite bank of the river with their family. Colyer explained that this suspicion led him to call-blast the laughter of children into the woods from time-to-time when primate calls were not getting results.

Alton Higgins recounted the events of Operation Thicket Probe when a large animal, believed to be a wood ape, was screaming and wheezing outside of the tent he was sharing with Colyer. "We had a number of vocalizations that were around our campsite. They were pretty dramatic vocalizations... close," Higgins said. "That was one of the most memorable experiences, I think, that any of us has had."[65] The concerns Higgins had voiced earlier— that viewers might think what he was recounting had occurred on the *MonsterQuest* trip—proved well-founded as his description of this event was not clearly identified as a recollection. Similarly, Colyer's recounting of the events leading up to the discovery of two tracks and a handprint was presented in such a way that a viewer would likely think that evidence was located in either Oklahoma or Arkansas when, in fact, it was discovered in the Sam Houston National Forest of southeast Texas many months previously.

Colyer was not happy with the experience. "If I were asked to do it again," he said, "I would decline. Speaking for the parts in which I was involved, there was no research done during the filming; it was all staged... the crew set up each scene; there was even a script of sorts as they would tell us the gist of what they wanted us to say and then we would ad-lib our lines according to the parameters they gave us."

Nonetheless the *MonsterQuest* experience was a positive one. Membership requests jumped, and the members were portrayed as earnest researchers.

Also, we all gained valuable insight into just how reality television works: there is not all that much reality involved in it at all. The positives that might result from an appearance on a legitimate program like *MonsterQuest* could potentially be dwarfed by the negatives associated with a creatively-edited piece that showed the group in a bad light. The issues and inconsistencies with the "Swamp Stalker" episode were annoying but nothing that anyone outside the group would likely even notice. Still, the dangers of having no editorial control were made clear. The group would need to be very selective about similar television projects in the future.

8
The Gulf Between Us

As Operation Forest Vigil was winding down in the fall of 2011, the TBRC began to experience serious internal strife for the first time. Like any group of more than just a handful of individuals, there had been disagreements and squabbles between members in the past; however, these had never been too serious in nature and were almost always resolved quickly. As the work of the group began to attract more attention, additional people sought to join its ranks. The additional manpower was welcome, but the group could no longer operate in the same casual manner it had in the past. For lack of a better term, the group had to become more *corporate* in nature. No longer could the TBRC be run as a "Mom and Pop-style" organization: a more formal approach was required. The Board of Directors drafted new documents and amended some older ones detailing the expectations of members. Among these documents were new versions of the group's bylaws, code of conduct, non-disclosure agreement, and a release of liability form. Membership was required to sign these documents and agree to abide by them. Some members balked at the new, and in some cases more stringent, standards for members.

The Board of Directors also sought to be regarded less as a "Bigfoot group" and more as a wildlife conservation group in the hopes that it would entice more scientifically-minded individuals to take a serious look at the work of the organization. In response to those concerns, a major rebranding took place in 2013, which included a new logo, website, and a new name: The North American Wood Ape Conservancy (NAWAC). These changes were not welcomed by all members. Some departed, but for the most part, even the members who were wary of the changes—or flat out disliked them—chose to stay.

The issue that caused the most dissension by far was the argument over whether or not the group should attempt to collect a type specimen of the wood ape in order to prove the existence of the species to science. The TBRC had always been a "no kill" group as the organization had never publicly advanced the idea of collecting a specimen. Leadership had always felt that high-quality photographs or DNA evidence collected in a non-lethal manner from hair or scat samples, for example, should be enough to prove the species was a real flesh and blood animal. The idea that suitably compelling photographs or video would suffice to establish the existence of the species had been the driving force behind Operation Forest Vigil. Some members had long doubted that a photograph or video would be enough and felt that a voucher specimen would be necessary, but they did not pursue collection efforts in the hopes that the group's camera project would get the job done. The failure of Operation Forest Vigil to provide conclusive evidence, coinciding with an influx of new people with different thoughts on the matter, led many in the membership to feel that a more aggressive approach was going to be necessary if the NAWAC was going to achieve its goal of documentation.

The debate became more and more heated. Those who thought the Sasquatch might be some kind of primitive human felt the collection of a specimen was barbaric and unnecessary, akin even to murder. Other members, including some who were wildlife biologists and familiar with what was needed to verify the existence of a new species, argued that there was simply no way around it; science demands a body. If the group was serious about its mission statement, which reads in part "…to facilitate scientific, official, and governmental recognition, conservation, and protection of the species…"[66] then procuring a specimen was the only way.

The Board of Directors, made up of seven longtime members, was split on the issue and found themselves presiding over a group that seemed close to fracturing. The Board decided to poll the membership, and somewhat surprisingly, the results showed that an overwhelming number of members supported the collection of a voucher specimen. Nonetheless, the Board was concerned with the possible exodus of valued long-term members who were against the idea. In an attempt to prevent their leaving the organization, NAWAC Chairman and wildlife biologist Alton Higgins issued a statement that basically said the group would officially take a neutral stance on the issue of specimen collection.

To those within the organization who were adamantly against the collection of a wood ape specimen, this declaration by Higgins was unacceptable, and the group did lose some valuable longtime members over this issue. One member who stayed likened the situation to a controlled burn in the forest. "It is painful to watch and leaves ugly scars, but the forest is left healthier for it. To achieve our goals, the group must be of the same mind."

At this point, I feel it is appropriate to delve into why the NAWAC eventually came to support the collection of a wood ape specimen. The topic is controversial, opinions on the subject are strongly held, and emotions run deep. It is not a conclusion the group came to lightly, but science demands a specimen. It was really just that simple. In order for the wood ape to be officially recognized, someone was going to have to bring in one of these animals. I realize there have been a few instances where science has recognized the existence of a species based on photos or video, but these cases are few and far between and never involved anything as controversial as a North American ape. To be honest, Bigfoot is regarded as on par with unicorns, dragons, and centaurs in the eyes of mainstream science. My fellow NAWAC members and I held out hope for a very long time that good video and/or photos would be enough but came to believe that was simply not the case. These animals have proven to be incredibly elusive, and sightings, on the rare occasions when they do occur, typically are so fleeting that even if a witness was holding a camera, he/she did not have time to get a good photo. Game cameras may ultimately get a picture, but will it be good enough? It's doubtful.

Besides, some high-quality images and film of putative wood apes have already been captured, and it has not been enough to sway the powers that be in mainstream science to officially recognize the species. The chances of capturing clear daylight footage more definitive than that shot by Roger Patterson in northern California in October of 1967 were thought to be extremely slim. This footage has been dissected in every conceivable way using modern techniques, and opinions on the authenticity of the film remain hotly debated to this day. The bottom line was that the footage was not considered convincing enough for mainstream science to recognize the species. The thinking was, if footage this clear was not good enough, no photo or video shot today would be either. If anything, it was my opinion that images today will be dismissed out of hand due to the proliferation of photo editing software available to the public.

Another reason the NAWAC came to believe a specimen was necessary was that—even if the species were to be accepted by mainstream science— mistakes could occur during conservation efforts due to false assumptions. Take, for example, the 2012 discovery of a new rodent in Sulawesi. It is unlike more than 2,000 other known species of rodents, as the *Paucidentomys vermidax* lacks cheek teeth, making it impossible for it to gnaw on its food.[67] This begged the question, what was this rodent subsisting on if it could not gnaw on nuts and seeds? "Stomach contents from a single specimen suggest that the species consumes only earthworms," the discoverers wrote. The scientists speculated that the species actually lost its gnawing incisors, which in most other rodents grow continuously, allowing it to "exploit resources that were not previously available, i.e. earthworms or other soft-tissued prey."[68] This crucial information could never have been deduced from photographs only. Luiz Rocha of the California Academy of Sciences summed it up well when he said, "Photographs and audio recordings can't tell you anything about such things as a species' diet, how and where it breeds, how quickly it grows, or its lifespan—information that is critical to assessing extinction risk."[69]

Opponents of the collection of a wood ape specimen sometimes recognize that photographic evidence will not be enough to convince science to officially list the species, and instead turn their hopes to DNA. Certainly, they argue, a DNA sample, obtained via non-lethal means, would do the trick. Unfortunately, that does not seem to be the case. Forget for a moment the difficulties of collecting such a sample, and think about how DNA testing works. DNA is sequenced and then compared to a database of known DNA samples. Simply put, if there is no match to a known species, and there is no type specimen from whence the anomalous DNA was taken, there is no documentation or recognition. The sample might be cataloged as *unknown*—something that has happened—or simply tossed out as contaminated. Either way, the DNA sample would be considered to have no scientific value, and the species would remain undocumented.

The NAWAC learned that a type specimen was a requirement in order for a taxonomic decision to be made regarding the species. Said David C. Howard at Centre for Ecology & Hydrology: "...I feel that the identification and characterization of species and their locations are essential; this will require the collection of a specimen."[70] Said Franco Andreone at the Museo Regionale di Scienze Naturali: "I feel that the simple use of photographs and tissue samples cannot provide a solid base for taxonomy."[71] Said Nigel P. Barker at the University of Pretoria: "To my mind this is a non-negotiable. Vouchers are essential for rigorous scientific practice, and I would reject any manuscript I review if there are no vouchers cited and detailed as to where they are housed."[72]

The importance of voucher specimens could not be underestimated. Photographs, video footage, or DNA extracted from hair or scat samples could serve as supporting evidence but can never adequately take the place of a holotype. To claim otherwise was, in my opinion to be a prisoner of emotion.

At this point, some may question the need to officially document the wood ape at all. Why not just leave them alone? While the sentiment is understandable, it is a recipe for extinction. Why? Habitat destruction. If the wood ape exists, then it is surely a rare animal. If we look at other primates, especially the great apes, it would seem safe to assume that wood apes are slow-growing and have low reproductive rates. If so, and these wood apes can only survive in the ever-dwindling heavily-forested remote areas of North America, then it's going to be in trouble very soon if deforestation and development continue unabated. One simply cannot overstate the effect deforestation is having on the planet's wildlife. It may be the greatest driver behind biodiversity loss. At one time, almost half of the continental United States and Canada were covered in forests. That is no longer the case.

Ninety percent of the virgin forest that once covered the lower 48 states has been cut and is not coming back.[73] This habitat is gone forever. Approximately 80% of the forest that remains is on public land. National forests, state forests, wilderness areas, state parks, and national parks contain most of what is left of our forests. Many would take comfort in that fact, but understand that this does not mean these wooded areas are safe. For example, approximately 80% of the forestland in the Pacific Northwest, the "Holy Land" of all things Bigfoot, is slated to be logged at some point in the future.[74] While logging has come a long way and is being done in a much more responsible manner than in years past, there remains little doubt that second-growth forests differ greatly in make-up from the old-growth forests they replace. It takes up to 100 years for a replanted forest to mature. Any species directly affected in a negative way by the original cutting of an old-growth forest will not be around 100 years later; it will be extinct. Perhaps the best example of this is how the logging of what became known as the "Singer Tract" in Louisiana was likely the final nail in the coffin for the Ivory-billed woodpecker.

The bottom line was that in order to protect the habitat of the wood apes, we first needed to prove to scientists and government officials that these animals are real. The government will simply not create specific areas that are off limits to logging and development for the protection of a mythical animal. Bringing in a specimen was the only way to prove, beyond a shadow of a doubt, that the wood ape is a flesh and blood animal. I just could not see how anyone who said they cared about the future of this species could make a cogent argument against the necessity of doing so. Labeling those who took this position as "blood-thirsty murderers" was not only mean-spirited but completely off base. There was not a single NAWAC member who wanted to take a wood ape as a trophy. They, like me, wanted to save them.

In addition, those who argued that the taking of a specimen would somehow accelerate the extinction timetable needed to realize that if the species was that depleted, it was already too late. American science, nature, and travel writer David Quammen summed it up best: "The crux of the matter…is not who or what kills the last individual. That final death reflects only a proximate cause. The ultimate cause, or causes, may be quite different. By the time the death of the last individual becomes imminent, a species has already lost too many battles in the war of survival…Its evolutionary adaptability is largely gone. Ecologically it has become moribund. Sheer chance, among other factors, is working against it. The toilet of destiny has been flushed."[75]

John Green, perhaps the greatest Bigfoot researcher of them all and author of the classic *Sasquatch: The Apes Among Us*, was also well aware of the stark reality of the situation: "What steps can be taken then, to ease man's pressure on a unique variety of wildlife? Obviously, none, as long as the authorities do not believe that any such creature is actually there. Where sasquatches are plentiful, the death of a few can be of no significance to the species. If in any area they are threatened by man's activities, it follows that something man is doing must be causing them to die or preventing them from reproducing, so he is already doing things that will eliminate them. If the first step towards changing the damaging activity requires killing one sasquatch to prove that there are such things, then obviously that is what should be done."[76]

But what if the Sasquatch is not an ape at all but some kind of primitive human; wouldn't killing one then be murder? To decide the matter is no simple task; we must answer what seems like a simple question: what exactly is it that makes us human? In 2016, *Live Science* published a brief list of "The Top 10 Things That Make Humans Special." The list includes characteristics like the use of speech, nakedness, use of clothing, the extraordinary human brain, and the mastery of fire.[77] When examined in detail, these characteristics clearly separate humans from wood apes.

John Green long ago came to the same conclusion: "There seems also to be a strong tendency for people whose fancy is caught by the idea of unknown giants living in the woods to want them to be human. The idea that there is a forest race too wise to have wars or get trapped in the rat race, living in harmony with its environment, seems to have tremendous appeal…[78] [But]… everything known about them proclaims that the sasquatch are not and never have been human."[79]

The members of the NAWAC, myself included, came to believe strongly that attempting to collect a specimen was the responsible thing to do in order to ensure the long-term survival of the species. One needed to be taken in order to save them all. We knew we could not wait; we had to do something. There would be no second chances.

PART 2:
Area X

"Once you have been to his land you may enter and leave at will, though few return from that journey unchanged."
— Margaret Atwood,
"Oratorio for Sasquatch, Man, and Two Androids"

9
A Land Called X

Area X. The name conjures up images of Fox Mulder and Dana Scully hunting down aliens and monsters while attempting to unravel government conspiracies. While Area X is every bit as strange and mysterious as the name suggests, that is not why the NAWAC gave this region of southeast Oklahoma the name. Area X was one of three study areas the group actively investigated when Operation Forest Vigil kicked off in 2006 until it wrapped up in 2011. These areas were nicknamed Areas X, Y, and Z (they could have just as easily been called Areas A, B, and C). At that point, the group decided that Area X would be its main area of study. Many in the group lamented the abandonment of the Big Thicket; the apes were there, after all, but there had been a lot of odd goings-on in the Ouachitas as well. In addition, Area X was privately owned land. The NAWAC would not have to continue the delicate dance required by the fickle National Park Service in order to get the necessary scientific research permits needed to conduct operations in the Big Thicket. Moving the focus to Area X seemed to be a no-brainer.

The NAWAC had discovered the amazing area now known as Area X back in 2001, when Alton Higgins and David Wilbanks of The Bigfoot Field Researchers Organization (BFRO) began exploring a mountainous region of southeastern Oklahoma in response to a number of reported sightings in the area. The most intriguing account came from a hunter who claimed to have seen a small, three-to-four-foot tall, chimp-like animal standing in a tree about 60 yards from his tree stand. When questioned by Higgins, the hunter could not bring himself to say he had seen a Sasquatch; "All I know," he said, "is that something is going on up there."[80] This report, among others, prompted Higgins and Wilbanks to start looking around the area.

After a frustrating day in which the pair discovered that the roads on the topographical maps Higgins had brought did not match up to actual roads in the region, the men ended up in a small combination convenience store/ grill south of Honobia. It was there that Wilbanks began actively questioning locals.

"Dave was so bold," recalled Higgins. "He had no reservations about asking, 'Have you ever seen a Bigfoot?' At first, it was embarrassing, but amazingly almost everyone he talked to had either seen one or knew someone who had and they would tell us their stories. It was amazing."

By chance, a man named Harold Yates walked into the establishment while the two investigators were visiting with the locals. Yates was a retired Forest Service employee who lived nearby. He invited the two men to his home where he revealed that, at least on occasion, he had apes on his property.

"Harold had a large wood shop on his property," Higgins said. "He built these really cool cabins and he milled the logs right there in his shop. He said that he thought the apes were attracted to the screeching sound produced by his saws as boards were being cut. He heard vocalizations that seemed to be mimicking the sound of the power saw."

Yates invited Higgins and Wilbanks to stay the night and during one of their talks Higgins asked: "Where would you go?" Without hesitation, Yates said "I'd go there," pointing to an area on Higgins's topographical map: an area that would later be known as Area X. "I can't get you in there," said Yates, "I don't know the way, but I can get you a guide."

Shortly thereafter, Yates introduced Higgins and Wilbanks to a man named Joe Teinert*. Teinert knew of an old road that led down into a valley in the area pointed out by Harold Yates and took them there. But calling the path that led down into that valley a "road" would actually be a gross exaggeration; the path was really only the remnants of an old wagon trail, full of ruts and jagged rocks that threatened all but the hardiest of off-road tires. The pitch was steep, and tree branches scraped against the roof and sides of Teinert's vehicle, making a screeching sound not unlike nails being dragged across a chalkboard. After more than an hour of bouncing, sliding, and weaving along the road and crossing multiple creeks that threatened to swamp the truck's engine, the men entered a small clearing where four primitive hunting cabins stood. They were arranged in a crude diamond shape, each one called by the cardinal direction they best represented.

That first night the three men spent a great deal of time at the north cabin as it sat very near the base of a mountain. One old adage to which most Bigfoot hunters ascribe is, whenever possible, cede the high ground to the apes. This places researchers at a tactical disadvantage but has proven effective in the past as the apes seem to feel more confident in this scenario and tend to approach closer.

Higgins and Wilbanks climbed up on the roof of the cabin to look for rocks and clear the forest debris. Higgins had made a habit of doing this over the years. "Rocks," he says, "do not drop out of trees. If they are on the roof of a structure, they have been thrown up there." By sweeping the debris and clearing the roof at the beginning of an operation, he knows that any rock found later in his stay had to have been recently deposited there.

Later, after the pair had retired to the south cabin—the largest and most secure of the four—the men began to hear something moving. "I think we had an ape visit us," Higgins said. "Something went to the east cabin and it was making a racket over there. Then it moved from the east cabin to the north cabin. You could hear it moving. Then we start to hear this racket again, like the place is being torn up. We then heard it continue to the west and into the woods." By the time all of this occurred, it was dark. The men, though immensely curious, were new to the area and thought better of attempting to follow the animal into an unfamiliar wilderness.

The next day, Wilbanks and Higgins checked the cabins for damage. "We couldn't find anything amiss," recalled Higgins. "We expected to find the biggest mess, but we couldn't see that anything had happened. It was the weirdest deal." The rest of the night passed quietly, and the pair left early the next morning. It was on the way out that one of the more important discoveries to come out of Area X was made.

The two men were slowly making their way up and out of the valley when they came across a low, flat spot made up of dried mud adjacent to the road. Such soft ground is rare in the rocky, thin soil of the Ouachitas, so the men decided to stop and scan the area for tracks. "We got out to look, and I saw this small concentration of leaves. I pulled the leaves out of what turned out to be a depression, and there is this track," Higgins said. "And it was big, like a 16-inch track." Wilbanks quickly located two more prints in the area that the NAWAC now refers to as the "mud hole." "That pretty much sealed it," said Higgins. "After hearing those noises and now seeing the trackway, I knew Harold Yates was on the right track."

Yates later set up a meeting between the investigators and the cabin owners. The family member who owned the cabins turned out to be very friendly and gave the men permission to stay in their cabins and conduct research. Plans were made, and an expedition was scheduled for September of 2001.

Higgins, along with four other BFRO investigators and one of the property owners, arrived on site on September 29. It was immediately clear that an animal of some kind had visited the south cabin recently. Something had managed to lift the heavy lid of a big chest-style freezer that sits outdoors, used for storing blankets and the like, and removed a large rolled mattress.[81]

NAWAC

Investigator Jerry Hestand pulling a casting of a large footprint from the banks of the main creek running through Area X.

Also, a piece of angle iron extending from the side of the cabin and support-ing a small solar-powered light was bent almost straight down. The property owner thought the culprit was likely a black bear, but the investigators were not so sure. They questioned whether a black bear would be so interested in inanimate objects that were not related in any way to food. Besides, the piece of bent angle iron projecting from the cabin wall was located at a height of seven feet. Could a black bear stand tall enough to reach that height? Later, Higgins spoke with Julie Davis, a BFRO investigator who had done field re-search on bears. She expressed doubt that the light fixture support could have been reached by any but the very largest of black bears. She admitted it was not an impossibility but added, "If you found a bear that size in the [Oklaho-ma study] area," said Davis, "it would be an important find."[82]

The men experienced some interesting activity over the next few days. On Monday, October 1, Higgins, who had chosen to spend the night in the woods a couple of hundred yards away from the owner's cabin, was awakened just before dawn by a sound he could not identify. "It was a sort of loud, low, drawn-out moan, unlike anything I've ever heard. It literally gave me shivers ..." A shaken Higgins went on to say, "I've spent hundreds of nights sleeping in remote areas, usually without a tent. Pack rats have crawled on me, skunks have tried to get into my sleeping bag, and I've heard all kinds of weird noises, most of which I eventually learned, but I've never had an involuntary response like I did that morning." The other team members, sleeping behind cabin walls approximately 200 yards away, did not hear it.

The men spent the day preparing for night operations. Pheromone chips impregnated with gorilla sex pheromones were deployed in the forest in several strategic spots. The researchers also set up a false camp, complete with tent, chairs, table, food, and a campfire, upstream from the cabin. The night proved to be mostly uneventful. Other than the sounds of coyotes, barred owls, and screech owls, the researchers reported hearing some distant sounds similar to alleged Bigfoot recordings, but they were too faint to record. They also reported being startled by what sounded like a rock forcefully impacting a tree at 10:00 p.m.

The next day, while some investigators set up a new audio recording site, Mike Simpson walked an additional mile to the north, hung up a pheromone chip, and then positioned himself in a tree stand overlooking the pungent bait. At 7:45 p.m., Simpson reported hearing the sound of a branch snapping. This snapping sound would be the start of the most harrowing event of the expedition. About 15 minutes later, Simpson radioed team members that he was hearing howls and screams emanating from a position to his southwest. Simpson would hear many other odd vocalizations and sounds before the night was over. Beginning at about 11:30 p.m., he heard the sounds of a heavy animal pacing back and forth and, at times, circling his location. The animal, whatever it was, would occasionally snap branches off of trees. He attempted to get a visual using his IR spotlight, but the vegetation around his location was just too thick.

Finally, the animal made a bolder move and seemed to be walking straight toward his location. Feeling quite uneasy, Simpson decided to bring his shotgun up to the stand; it had been hanging below the stand via a tether. That apparently gave the animal pause, and it stopped. For several minutes, there was no sound at all. Could it see him? Did it know the gun represented danger? After 10-15 minutes, the animal resumed its previous activity of loudly walking around his location while occasionally snapping off tree branches.

Simpson had had enough and radioed the team at base camp to ask for an escort out of the area. The sound of Simpson's voice seemed to disturb the animal, and he heard it slowly walk away. As the other two BFRO investigators on the trip, Roger Roberts and Brett Elliott, approached Simpson's location, they heard the creature bolt. It had not gone away at all; rather, it had positioned itself on the trail just south of Simpson's location. (The question of whether or not it might have been lying in wait for Simpson was the subject of much discussion later.) The pair froze upon hearing the creature bolt. Initially quite frightened, they thought the animal was running toward them. Instead, the creature was running away from them, passing Simpson's location using a nearby creek bed as its escape route. "I could hear it splashing," Simpson said later. "It covered more rough area quicker than I can conceive."[83]

The operation continued for several more days without any further intense activity, so the team packed up and left the valley, now firmly convinced it was home to at least one Sasquatch.

The next year, a small team consisting of Alton Higgins, two wildlife biologists from the Idabel, Oklahoma, area, Roger Roberts, and several of the landowner's family members returned to the property. The group was sitting outside the first night and decided to do some call-blasting. The family, in particular, was intrigued to see what, if any, responses the broadcasts might generate. Roberts blasted the sound of the Ohio Howl multiple times. The eerie, moaning call echoed up and down the valley, but nothing responded. Roberts decided to change tactics and played a putative Sasquatch vocalization called the Tahoe Scream (due to it having been recorded south of Lake Tahoe, California). There was an instant reply.

"It was exactly like what had been broadcast," Higgins recalled. "Exactly the same, and it was close. It was much louder than what Roger had blasted with his sound system. It was stunning." The group was completely taken aback. Nobody present, including the wildlife biologists, had ever heard anything like it before.

It is because of these early experiences in southeast Oklahoma that Area X was selected as one of the camera sites for Operation Forest Vigil. Strange things—rock throws, vocalizations, bangs on the sides of the cabins, and more—were experienced by members who went to the valley to tend to game cameras. These events led the NAWAC to make Area X its primary study area (despite the fact that the majority of the organization's members lived in Texas).

10

The Ouachita Project

Alton Higgins was now convinced that there was an unlisted species of primate living in this remote area of the Ouachita Mountains. His resolve was only strengthened when he made another trip to the site with new group member, Daryl Colyer. The trip was eventful. The group, which also included Colyer's wife and step-son, documented nightly rock-strikes on the cabin in which they were staying and located a set of very fresh tracks on wet mossy rocks, complete with apparent toe impressions, spanning up the embankment of a nearby creek.[84] And so it was that the valley, now dubbed Area X, was made the focus of the group's long-term camera project, Operation Forest Vigil, which kicked off in 2006.

Over the life of Operation Forest Vigil, multiple group members made trips into the valley, and many of them experienced activity that seemed likely to be attributable to the target species. In 2006, Alton Higgins and Ken Helmer reported having a large rock thrown at them as they hiked. "It was my first trip into Area X," Helmer recalls. "We were hiking the trail up the north mountain when I heard something clipping vegetation. A moment later, a rock landed about 20 yards to our right. I said to Alton, 'I think something just threw a rock at us.' He replied, 'I think so, too.'"

This rock-throwing incident took place on the second to last night of a week-long stay in the valley. On the final night, Helmer and others on his team reported hearing some kind of wood impact sounds during the night. The sounds seemed to originate from two separate locations about 100 yards apart and just south of the cabin. Helmer said, "They were sharp cracking sounds that might best be described as loud knocks on wood or trees…" It should be noted that no one in the group was yet convinced that "wood-knocking" was a legitimate part of the wood ape phenomenon. While experience has come to change the minds of everyone in the NAWAC regarding this behavior, the team—while agreeing it was an extremely odd thing to hear in the middle of the night—was unconvinced the sounds were of any significance regarding their mission.

Over the next five years, multiple group members braved the treacherous road into Area X in order to refresh batteries in the game cameras on site and download images. No conclusive images of wood apes were captured during the Operation Forest Vigil years, but team members began to recognize a certain pattern that seemed to repeat itself on each trip into Area X. More often than not, the first few days in the valley would be quiet with little to no suspicious activity. But toward the end of a team's stay—often on the very last night—suspected ape activity would spike considerably. There was often banging on the walls of the cabins where the investigators were staying. Sometimes it sounded like a thrown rock; sometimes it was more like the slapping of a hand against the wall. Other sounds often heard toward the end of a week included "whoop" vocalizations, long screams or howls, clacking sounds—as if two rocks were being banged together—and an increasing number of wood-knocks.

Why the apes only start acting out toward the end of an operation was something we had discussed many times over the years. A theory evolved regarding this pattern of activity. I wish I could recall exactly who first proposed it so succinctly, but I cannot. The gist of the hypothesis is that apes have been watching humans for a very long time, and while they do not like having their territory invaded by us, they have learned that people typically come in for a night or two and then leave. But perhaps after several days of human presence the apes lose patience and attempt to run the intruders out of the area. Activity is often subtle at first, then escalates until the people are frightened off. Think about it, most people would flee very quickly after having rocks thrown at them, hearing screams or howls in close proximity, or other noises such as loud wood-knocks or branches breaking. Perhaps the reason activity seemed to occur only at the tail-end of NAWAC operations was because, for lack of a better way to put it, we had begun to get on the nerves of the resident apes.

Similar odd activity had been documented by family members of the land owners in logbooks as far back as 1999. The summer months were times of peak activity. This summer spike was noted by Higgins, Colyer, and Helmer and was given much consideration when discussing what would follow after the conclusion of Operation Forest Vigil.

In November of 2010, Higgins, Colyer, and Helmer returned to Area X for a weeklong stay. The trip would prove to be a turning point of sorts for the NAWAC regarding how the group would go about attempting to document the wood ape. Members were also forced to reassess their opinion on wood-knocks.

"I've been going to Area X for ten years," wrote Higgins in an email upon returning home. "I've been there dozens of times. I've heard vocalizations, found the best tracks I've ever seen, discovered hand prints on vehicles, talked with folks who have seen wood apes, and experienced other forms of possible (in my mind probable) wood ape behavioral manifestations. One thing I haven't come up against is "wood-knock" sounds. Until this week.

"Honestly, as Daryl can attest, I've always been rather dubious of wood-knocks as a sign of ape activity. This week I and others heard loud and clear what I can describe as wood-knock sounds. We heard them in the daytimes and at night. Two that I heard seemed related, with a louder first knock and what appeared to be a reply with a duller sound coming from a farther distance. So, I still don't quite know what to make of the whole deal…"

Other team members heard the strange and enigmatic knocks, including Brad McAndrews* who wrote: "Late that night somewhere between 11:30 p.m. and 1:00 a.m., Ken Helmer and I heard at least two or three clearly defined wood-knocks from behind the east cabin and, again, up on the mountain to the north. One of the knocks has left me dumbfounded. As we sat around the warmth of our fire pit, we heard an interestingly 'clean' and seemingly close wood-knock. The reason I feel, a bit dumbfounded is because of a couple of characteristics. First, the sound was extremely clean, close, and 'soft,' as if not to make too loud a noise. (That was honestly the impression that I had) Secondly, this was a dual-knock, meaning that there were two identical sounding knocks one after another with only half-a-second in between the two. The odd thing, though, was the tone of the wood-knock. It sounds almost exactly like a musical instrument…the wood block." (See sample sounds in the Sound File Appendix.)

McAndrews was unnerved by the wood-knocks as they seemed so close and intentional. Said McAndrews, "If you think about this idea, that a wood ape was there among us communicating in an intentionally subtle way to other wood apes, it's a hard pill to swallow. If true, it's no wonder why the species has not been documented."

Higgins, too, struggled with the phenomenon and what it might mean. "As I've stated," he said, "it's still hard for me to wrap my head around something that I've tended to dismiss. Now I've suddenly been confronted with it, and it's an uncomfortable intellectual position."

The NAWAC leadership agreed that in the aftermath of Operation Forest Vigil, a new and bolder approach to research was in order. The membership seemed to be chomping at the bit to get into Area X and experience these strange events for themselves. Members contemplated not only what the purpose of the odd knocking behavior might be but also new ideas on how best to get to the bottom of it all. It was Travis Lawrence, a math teacher from Splendora, Texas, who gave the discussions some specificity when he suggested the group might be able to pull off a long-term field project in Area X.

"I propose," Lawrence wrote in an email to group leadership, "that over the following summer we conduct an extended research project in Area X. To my knowledge, the longest we've ever stayed at those cabins continuously was five to six days. I think we could realistically have people in Area X for at least 30 days straight, if we plan well. Instead of ten people going to X for four to five days, perhaps those ten people could go in groups of two for five days each, resulting in 25 days of research. Plus, small groups of people may tend to experience more contact."[85]

Field Operations Coordinator Daryl Colyer then expanded on Lawrence's suggested plan and proposed a six-week project. Teams would be composed of no more than three or four individuals who would stay in the valley for a week until their relief arrived. If all went according to plan, the NAWAC would have a continuous presence in Area X for 42 consecutive days.

What might happen if we acted in an atypical manner and did not flee the area when the apes began to show their impatience? If we held our ground and did not leave, would their behaviors escalate to the point where an ape might reveal itself? These were the questions the group wanted to answer with its long-term field study project. The study was scheduled for the summer of 2011, and the Ouachita Project was born.

11
Operation Endurance

All the information that follows about the experiences of the members of the NAWAC over the last decade in Area X is drawn directly from the field journals of the various participating teams, their compiled notes or after-action reports (AARs), the *Ouachita Project Monograph*, and personal interviews. In drawing from these sources to put together this book, I was taken aback by the sheer volume of data the NAWAC has collected over the years. There is simply no way I could include everything that the NAWAC has experienced in Area X. Missing are literally hundreds of "minor" events like anomalous wood-knocks or rock-throws. I had to limit the events detailed here to those I deemed the most significant, while still capturing some of the feel of the place during day-to-day activities.

While the overall field study in Area X was christened the Ouachita Project, each individual summer operation was given a name as well. Operation Endurance took place in the summer of 2011. I had the privilege of kicking off Operation Endurance as a member of Alpha Team on June 4, along with Field Operations Coordinator Daryl Colyer. We arrived at the cabin compound in the late afternoon.

That night, Colyer awoke to the sound of something striking the window screen above his bunk in the second story loft of the south cabin. We were awakened at approximately 3:00 a.m. and again at 5:00 a.m. by the screams of some kind of animal to our south. There was much movement in the brush along the creek from that direction as well. We both agreed the screams had likely been produced by a gray fox but also realized that whatever we were hearing moving in the brush was a much larger animal. Later, at 6:10 a.m., I was awakened by the sound of knocking. I heard six clear knocks that were spaced one to two seconds apart. The sound had a definite wood-on-wood tone.

In the morning, we deployed eight trail cameras: four Reconyx Hyperfires, three RC-55s, and one RC-60. These were the best trail cameras commercially available at the time. We then went on a cross country hike through the woods to the southeast of the cabin. Not far from a small tributary of the main creek that cuts through the area, we located a long bipedal trackway. The prints were old and somewhat degraded but were clearly made by a biped of some kind. The prints were not exceptionally large, just 12 inches in length, so we could not rule out a human as the culprit. Still, we were encouraged by the find.

The next day, Travis Lawrence and Tod Pinkerton, both veterans of the Big Thicket days, arrived. That evening a large animal of some kind was heard moving through the brush north of the cabin. No other activity was noted before we turned in for the evening. But at some point during the night, Colyer and I were both awakened by an odd "Oooaaahhh" vocalization from southwest of the cabin. The call started low and ended on a higher note. The two of us heard the vocalization on three different occasions during the night. The call was repeated between three and six times during each event. I remember asking Colyer, "Just what the hell kind of place is this?"

The following evening, we had an experience that would confirm what we had long suspected about our trail cameras: they did not work very well. At 8:55 p.m. we were gathered on the porch of the two-story cabin when we heard a doe blow off to our east. Shortly afterwards, a fawn bolted out of the woods near the creek and wobbled across the trail to the brush on the other side. Mere minutes later, a black bear emerged from the same spot and made his way on to the trail. The bear, which we estimated at 200 pounds, sniffed the ground intently and did not seem to notice us at all. After a minute or two, the bear did catch our scent and looked right at us. A few moments passed before he went back to sniffing the ground and sauntered into the woods on the other side of the trail, we assumed, in pursuit of the fawn. We were excited as the bear had spent a solid two to three minutes in front of no less than three of the game cameras we had deployed. We knew, if nothing else, we would have some great bear photographs to share with the membership once we returned home.

At 9:00 a.m. the next day I was awakened by a series of six clangs coming from somewhere west of our location. The "clangs" were rhythmic and had a metallic sound as if someone was striking a rock on a fencepost or the metal property gate. I alerted Colyer and went to open an upstairs door that faced to the west and immediately heard another clang from less than 100 yards. We dressed and checked out the area to the west but found nothing.

Just before lunchtime, we decided to pull the cards from the cameras and take a look at the bear pictures we were sure had been captured the previous evening. We were stunned to find that not a single camera had fired during the event; we had not one bear photo. We could not believe it. The bear had been in the *exact* spot where three cameras had been aimed. The batteries were fine, and the cameras were not mounted too high. We now had definitive proof that even the best cameras out there were not very reliable at all.

Later that evening, Pinkerton and Lawrence deployed in a ground blind to the east of the compound. At 6:45 p.m., the pair heard a strange, multi-tonal vocalization. "It was made up of several 'notes,'" Lawrence said, "as if forming a chord. It was like a choir singing out there in the middle of the woods."

At 10:00 a.m. on the morning of June 10, Lawrence walked over to the north cabin, which sits at the base of one of the two mountains that frame the valley. Within a second or two of closing the door of the north cabin's outhouse, Lawrence heard a loud impact on the roof or one of the walls of the cabin a few yards away. "It was much too loud to have been a falling nut or a sweetgum ball," said Lawrence. Over at the south cabin, Pinkerton heard the impact and made his way over to the north cabin to make sure Lawrence was okay.

After discussing the matter, the pair decided to recreate the scenario leading up to the impact. Pinkerton positioned himself in the patch of woods between the north and south cabins and observed while Lawrence walked back to the outhouse, entered it, and closed the door. No more than 30 seconds later, something else struck the roof or back wall of the cabin. "It was really, really loud," recalled Pinkerton. Lawrence sprang from the outhouse, hit a knee, and scanned the mountain slope behind the cabin. He couldn't see anything at all on the mountain. Neither could Pinkerton.

Over the next 36 hours, Colyer, Pinkerton, Lawrence, and I heard multiple wood-knocks, bangs on the property gate, rock-strikes on the roof of the north cabin, a large tree fall on the mountain slope, and a whoop vocalization from the north. We agreed that Operation Endurance was off to a good start.

On Saturday, June 11, the Bravo Team arrived; it consisted of Bill Coffman, Mark McClurkan, and Travis Lawrence, who had decided to stay on for an additional week. The second week of Operation Endurance would also prove to be eventful.

Around lunchtime the next day, the men heard a white-tailed fawn in distress. "It was afraid," McClurkan said. "It sounded like it was bleating for its life." Shortly after the fawn began vocalizing, the men heard a very loud "thump." This seemed to trigger the mother doe who began to blow in a panicked fashion. Lawrence grabbed his rifle and began to scan the area east of the cabin. The alarmed fawn emerged from the woods that bordered the creek and on to the trail that led to the east cabin. Strangely, it began to run toward the men. Lawrence took this as a sure sign a predator was close behind and trained his rifle on the trail behind the fleeing fawn. Minutes passed but no predator emerged. The doe "continued to blow, but was getting more and more panicked," McClurkan wrote in his field notes. "I think she was trying to confuse the predator, whatever it was, and protect her fawn."

Lawrence and McClurkan decided to follow the fawn down the trail to the west, figuring the predator might have traveled west down the creek bed behind the cabin (instead of coming down the trail in plain view of the men as the terrified fawn had done) in an effort to avoid detection and cut off the young deer. Suddenly they heard a menacing growl from the woods ahead. "It was a low growl," Lawrence said. "It was deep, guttural, and threatening."

The men peeled back to the northeast and walked toward the north cabin, where they encountered a terrible odor. "It was malodorous," Lawrence wrote in his field journal, "a dead animal smell. A primal, nasty smell. It lasted approximately one minute, then it was gone." The team reported that the doe continued blowing off and on for the rest of the evening. They could only speculate as to what became of the fawn.

The evening was active and the men documented multiple strange sounds, including knocks, loud impacts on the roof of the north cabin, and a "huge snapping sound" followed by the sound of a large tree crashing to the ground. The team was totally enthralled by the cacophony of unusual sounds, which would soon intensify.

Shortly after 11:00 p.m., McClurkan, a seasoned outdoorsman, was sitting alone on the south side of the porch near the creek (everyone else was inside the cabin) when he heard a guttural growl close by. Unnerved, he moved his chair to the far north end of the porch away from the creek and the growler. Moments later, he detected a strong odor drifting in from the creek bed to the south. "It smelled like sour sweat from a horse after the saddle has been removed at the end of a long ride," he said.

Around 1:35 a.m., Lawrence, McClurkan, and Coffman were on the porch when they heard what they described as "clear bipedal footfalls to the south of the cabin along the creek," which would put the walker mere feet from the cabin wall. "We could hear the rocks grinding together underneath its feet," Lawrence wrote, "and we heard no concussive pop like what would be heard of someone wearing shoes. This was a heavy animal with padded feet that was walking on two legs." Even using the night vision monocular, the team could not get a visual of the walker; the foliage was just too thick.

Knowing the animal was uncomfortably close, they decided to retreat inside the cabin. A few minutes later, at about 1:50 a.m., McClurkan opened a west-facing first floor door in an effort to get a glimpse of the visitor. "I heard the bipedal footfalls as soon as I opened the door," McClurkan said. "The animal then broke into a sprint. It sounded like it had run across the trail and into the forest for cover." The men lit the area up with white lights, but saw nothing.

Soon after, the men climbed up to the second story loft in hopes of getting some rest. Sleep was hard to come by, however, as the night continued to be filled with strange sounds. Around 2:45 a.m., the men again heard "clear bipedal footsteps" to the south. They listened in silence for several minutes. The animal seemed to be pacing in a semi-circular pattern to the north and then back to the south toward the creek. A few minutes later, the men heard a sound that was determined to be the barbed wire fence (to the west of the cabin) being compressed. "It was as if the animal was pushing the fence down in order to step over it," said McClurkan. The rest of the night was filled with banging sounds from the north cabin, more bipedal footsteps, and a large crash that the team speculated was caused by a tree-fall.

The men reported more activity in the early morning hours of Thursday, June 16. At 1:27 a.m., McClurkan produced three wood-knocks, which were answered with a loud "whoop" sound from the north. Less than an hour later, the men heard an incredibly loud bang from the area of the east cabin. The sound was described in the team's field notes as, "big, big, BIG!"

At 2:40 a.m. the team described a rather incredible event, which was recounted in their field notes: "We all heard a series of definite wood-knocks. The first knock came from the same vicinity of the 'whoop' that had been in response to McClurkan. Perhaps one or two seconds later, the team heard a double wood-knock from the north-northeast perhaps a few hundred yards away. One to two seconds later, we heard another closer double wood-knock from the northeast by the shed. A few seconds later came a single wood-knock to the south of our cabin from a few hundred yards away. We were dumbfounded as the knocks—all in response to one another—happened within a span of ten seconds. We believe there were four wood apes in and around the encampment area."

The remainder of the night was filled with metallic bangs, the sounds of breaking limbs, odd sweet smells, and one "almost musical-sounding" wood-knock. Exhausted, the team turned in for the night at 5:45 a.m.

At noon the next day, Bill Coffman left the valley to return home. He had traveled only a mile or so to the west of the cabins when a large, hair-covered animal bolted across the road in front of him. The creature was brown and ran bipedally. Coffman observed muscular striations on the back of the animal as it leaped across a creek, as well as a lighter-colored "stripe" running down the mid-line of the back, before it disappeared into the thick woods. Coffman was stunned, but instead of returning to camp, since places to turn around are few and far between on the narrow road and backing up for any length of time is not feasible, Coffman made his way home and reported the visual to group leadership.

The day and night of June 18 was filled with the sounds of a large animal thrashing about in the woods near the cabin, intimidating growls, periodic foul odors, one faint but long "Ohio Howl-like" vocalization, and, according to Lawrence, "easily the loudest sound heard thus far"—a bang at the east cabin.

After a full day, McClurkan and Lawrence retired for the night at 4:30 a.m. Sleep, however, would not be in the cards. Only ten minutes later, they heard the sounds of a large animal moving from the south to the northeast. After which an overpowering, putrid, and pungent "urine smell" floated all the way up to the loft. At approximately 4:45 a.m., the men heard a large animal of some kind move up the creek from the south, compress the barbed wire fence, and "jog" away. Within ten minutes, McClurkan and Lawrence heard what they described as large, bipedal animals moving on all four sides of the cabin. During the event, the smell of rancid urine intensified.

Just as first light was beginning to soften the inky darkness, the pair of investigators experienced what Lawrence described as "the single most bizarre event of my life." The men heard what they described as a "squishing sound" below them on the north side of the cabin. "I'm certain," said Lawrence, "the sound came from the mouth of some animal that was gurgling a copious amount of liquid or saliva, and it was loud." Then another animal below them (on the south side of the cabin) made an identical sound. The first animal continued to produce the bizarre sound even after the second creature joined in or answered. Amazingly, a third animal then joined the squishy chorus from the northwest.

McClurkan and Lawrence were positively stupefied; the exact same squishing sound was being produced by three individual creatures on three different sides of the cabin. Each animal produced the sound continuously every two or three seconds for several minutes. Finally, the men heard two of the animals make their way to each other and walk away to the north, continuing to produce the strange sound even as they retreated. The third animal stopped producing the sound but it was unclear whether this individual had left the vicinity of the cabin or had simply gone silent.

While discussing what had just occurred, the investigators were interrupted by deep, guttural growling below the loft on the south side of the cabin. It seems the third creature had not left the area after all. McClurkan attempted to reproduce the squishing sounds the men had just heard. The animal below, hidden in the wood line, growled even more menacingly in response. This went on for several minutes until an exhausted Lawrence yelled, "Are you just gonna sit down there and growl at us all morning?" The animal then went silent and the urine smell, which had been oppressive during the entire encounter, disappeared as well.

Shortly before 6:00 a.m., the two men heard yet another bizarre vocalization. The sound was multi-tonal and repeated several times. Lawrence said it was akin to "a pipe organ," like an "ahhhhh-ohhhhh" that started with a low note and descended to a lower note not unlike a foghorn. "No one will believe this," McClurkan muttered.

Lawrence, who had been on site for two weeks, and McClurkan were absolutely exhausted, and their nerves were shot. After a few hours of desperately needed sleep, they began loading their vehicles and getting ready to leave. While doing so, they heard "two clear Ohio Howl-type vocalizations" emanating from the top of the mountain to the north. Ten minutes later, two more long, moaning howls were heard. Both men marveled that they were hearing this type of vocalization during daylight hours.

Finally, just before leaving the valley, the men reported hearing the loudest sound of the week. "It was like the sound of a big rock falling or being dropped onto another big rock," said McClurkan. "Upon impact, it sounded as though one or both of the rocks split." The sound, Lawrence wrote in his journal, was like that of "dynamite." The men could not adequately describe the sheer volume of the event and could not discern how close or far away it was. "It was the most unnerving event of my two-week stay," Lawrence said. "I find it unbelievable that any animal could have produced a sound of such magnitude." The two exhausted men then left the valley and met up with incoming Charlie Team members Phil Burrows and Alton Higgins on the morning of the 18th.

After unloading, Burrows and Higgins examined the roofs of several cabins in the hopes of finding rocks. They then strung black thread between trees located along three suspected travel corridors at a height of approximately six feet. The thread was of the strong industrial variety and was tied securely to one tree while the other end was wrapped—not tied—to another tree. The string proved to be all but invisible in the densely wooded Area X environment. The idea behind these "string traps" was that only an animal in excess of six feet in height would walk through and disturb the string, likely feeling nothing. The wrapped end of the string would come undone and stretch out in the direction the animal was traveling, providing valuable data on movement patterns.

Between 8:00 and 9:00 p.m. on the 19th, the men produced several wood-knocks by striking a tree with a wooden bat I had left behind. While there were no immediate responses to their efforts, the men heard a loud knock from the east at 11:00 p.m. The night was still and calm, and both had no doubt that something or someone had deliberately manufactured the sound. "This knock was easily the most dramatic and distinctive I have heard," said Higgins, who had only heard wood-knock-like sounds once before in this location. The men responded with a knock of their own but received no reply. The next morning the pair were greatly relieved to find that the knocks had been successfully recorded. "No Bigfoot curse today!" Burrows remarked.

Shortly before 3:00 p.m. on the afternoon of the 20th, investigators Jerry Hestand and Chris Buntenbah arrived in the valley. They informed Higgins and Burrows that another team would be entering the valley late Friday night or in the wee hours of Saturday morning. The plan, dubbed Operation Flanker, would involve the members of Delta Team—Daryl Colyer, Paul Bowman, and Jeff Davidson—entering the valley on foot as stealthily as possible. Would any resident apes observing the group of men at the cabin be caught unaware by the quietly advancing Delta Team?

At approximately 8:15 p.m., Jerry Hestand headed to an area on the slope of the north mountain and positioned his tree stand 25 feet up in a stout hickory. He was fully camouflaged, equipped with night vision, a 12-gauge shotgun, and white lights. Meanwhile, Buntenbah deployed one of the group's temperamental mini-disc recorders. The night would prove eventful in short order.

At 11:21 p.m., the team heard an extremely loud noise like a rock striking metal from the direction of the north mountain. Only minutes later, they heard another tremendously loud sound. "It was not like before," said Burrows, "it sounded like something was actually striking the walls of the north cabin itself." At midnight, Higgins heard the sound of a rock hitting metal from the area near the north cabin. "It was LOUD," Higgins wrote in his notes. The remainder of the night was filled with the sounds of rocks striking metal, wood-knocks, and the footfalls of an unseen animal moving through the brush and along the creek. The activity was so constant that Higgins had trouble recording it all.

At approximately 2:00 a.m., Hestand returned to the south cabin. He confirmed he had heard all of the noisy activity from his perch near the north cabin. He also reported that a stick or broken limb had whizzed "end over end" in close proximity to his head as he sat in his stand. Angered, Hestand initially thought the group might be the victim of hoaxers. "Hillbillies are jacking with us!" he said upon his return to the cabin. The team managed to calm Hestand down and convinced him that it made little to no sense to think hillbillies were roaming around in the tar black night playing pranks on them in this extremely remote location.

The next morning, Higgins noted in his journal that "things got pretty crazy for a while last night." At 10:00 a.m., Buntenbah went to check on the mini-disc recorder and found that the balky device had failed to record any activity. The Bigfoot curse appeared to be alive and well after all. Later, Burrows climbed to the top of the north cabin—the source of so much activity the night before—and located a rock measuring 1 x 2 x 3 inches. Team members did not have to be reminded that rocks do not just fall from the sky or fly up onto rooftops.

Around 10:00 a.m. on Friday, June 24, Hestand, Burrows, and Higgins located a new spot for tree stand deployment, one just below the east cabin that would provide a good view of the creek bottom in two directions. The investigators then went to check on the various cameras that had been deployed the day before. The men soon discovered that a camera on a post near the south cabin had been tampered with. The movement was recorded at 8:52 p.m. the previous night while the team had been present at the cabin, mere yards away. The movement had caused the camera to fire multiple times. A second Hyperfire camera, which had been placed for the express purpose of observing the camera on the post, failed to take a single photo. "Frustrated with cameras!!" Higgins wrote in his field notes. Chastened, but unwilling to give up, the men redeployed several cameras in an elaborate "camera-watching-camera" manner in the hopes of capturing the "camera-shaker" in the act.

At 2:40 a.m. on the morning of the 25th, Delta Team—Colyer, Bowman, and Davidson—arrived at the property gate. Leaving their vehicles at the top of the mountain, the men had walked the road to the compound in complete darkness hoping to take any resident apes by surprise. Shortly after arriving, the men banged on the property gate five times with a rock. They received no response and continued toward the cabins. They had walked only 30 yards when they flushed out a large animal of some kind to their left. The animal fled into thick vegetation in the direction of the north mountain and was never visually acquired. At 3:30 a.m., the men arrived at the south cabin and met up with the members of Charlie Team. Other than two faint wood-knocks, the remainder of the night was quiet.

At 9:00 a.m., most of the team drove up the mountain in order to retrieve Delta Team's gear, while Higgins and Colyer walked to the east on a reconnaissance. About 100 yards from the eastern-most cabin, Higgins found a piece of cut and split firewood lying on the ground next to a small tree. The men guessed it had been snatched from the porch of the east cabin at some point. Higgins picked up the firewood and struck the small tree with it; the sound produced was identical to the knocking sounds Charlie Team had heard from that vicinity the previous week. The men moved on to the north cabin and checked the metal roof for rocks. While no rocks were found on the cabin roof, Colyer did locate a flat rock about two inches in diameter on top of a nearby metal shed. The rock had to have been recently thrown there as Higgins had done a sweep of the cabin and shed roofs upon arrival a week earlier.

Travis Lawrence and Ken Helmer, the first members of Echo Team, arrived at the property gate at 10:00 p.m. on Wednesday, June 29. When Lawrence stepped out of his vehicle to open the gate, he was greeted by a familiar pungent odor. The smell was strong, indicating that an ape might be in close proximity. After settling in, the two investigators spent the next several hours sitting on the cabin porch. Lawrence noted in his field notes that "The smell comes and goes. When it is present it is very strong; stronger than any smell I encountered during weeks one and two."

Shortly after midnight on Saturday, July 2, team members Alton Higgins, Daryl Colyer, and Alex Diaz decided to patrol the compound and try out their new TIM-14 thermal imaging monocular provided by investigator Paul Bowman. The team was impressed with the ability of the new thermal but saw nothing of consequence.

At 11:00 a.m. the next morning, Higgins, Diaz, Helmer, and Lawrence went to inspect the roof of the north cabin. The cabin, which sits at the base of the north mountain, had continued to be the origin of multiple impact sounds over the previous few nights and Higgins wanted to inspect the roof for rocks again. He would not be disappointed; there were multiple stones on the corrugated metal roof. The roof had been cleared of debris back on the 18th of June, meaning these rocks had somehow made their way onto the cabin over the last couple of weeks.

Baseball-sized rock on cabin roof.

The act of walking on the metal roof was quite noisy and the corrugated metal popped loudly with each step Higgins took. As he did so, the men heard an "extremely peculiar vocalization" coming from about 150 feet up the slope to the north. All agreed it was as if some animal was attempting to mimic the sound of the pops and groans of the buckling metal as Higgins traversed the roof of the cabin. The team found the sounds perplexing and "more than a bit unnerving."

At 11:30 p.m., Colyer and Lawrence returned to the south cabin. As soon as they stepped up onto the porch they heard a loud, clear, and close wood-knock emanating from somewhere between the north and east cabins. Colyer and Lawrence made a beeline for the "firewood tree" where Higgins had earlier located the piece of cut and split wood. The piece of firewood was still at the base of the tree and Higgins used it to strike the tree; the sound produced was identical to the knock they had just heard. After producing multiple knocks, the pair of investigators heard the noise of a large animal moving through the woods to their west. The men rushed to investigate, but despite hearing the sound of running footfalls nearby, they could not visually acquire their target. Colyer and Lawrence continued to search, but eventually gave up and returned to the south cabin, chastened and irritated that, once again, they had been manipulated.

Shortly after 11:00 a.m. the next day, the group heard several rock-strikes on the roof of the north cabin. The men investigated but found nothing of interest. They all returned to the south cabin except for Lawrence who remained concealed in the woods. For the next two hours Lawrence observed the north cabin. At 1:30 p.m. he decided he had watched long enough and stood up to leave. He had scarcely taken a step when a rock struck the cabin. Frustrated and angry, Lawrence sprinted to the side of the cabin, picked up a large rock, and hurled it with all of his might up the slope, letting out a loud "Ahhhh!" as he did so.

The thrown rock struck a large boulder embedded on the mountain slope and split loudly. Lawrence yelled up the slope: "You're not the only ones who can throw rocks around here!" Alarmed by the sound of Lawrence yelling and the loud rock-on-rock sound, the rest of the team rushed to his location. After getting the story from Lawrence, the men—who completely understood his frustration—shared a good chuckle at their predicament. Little did they know that in the matter of just a few hours, one of the most important events in group history would take place, an event that would become known as the Echo Incident.

12
The Echo Incident

The events that transpired on the afternoon and evening of Sunday, July 3, 2011, became known simply as "The Echo Incident." It is one of the seminal events in the history of the NAWAC, and it remains a controversial topic among members of the Bigfoot community. The misinformation about the incident runs the gamut from misunderstandings, to half-truths, to outright lies. For someone who has come to hear of the incident after the fact, it would be difficult—if not impossible—to discern exactly where the truth lies. Here are the facts—what occurred before, during, and after the incident, as well as the fallout in the aftermath— so as to end the speculation and misinformation that still swirl around this remarkable episode.

By 6:00 p.m., the team had eaten their dinner and were discussing what sort of activities they might undertake that night. Suddenly they heard a loud impact sound—like a rock striking metal—from the direction of the west cabin. Colyer, Lawrence, and Diaz all moved quickly to investigate, as they felt confident that what they had heard had been a rock and not a falling hickory nut. Colyer grabbed his semi-automatic shotgun—loaded with two initial rounds of .00 buckshot followed by seven shells containing rifled slugs—and began moving down the ATV trail to the west toward the cabin. Diaz trailed him and maintained a distance of about 25 yards from his teammate. Meanwhile, Travis Lawrence split off from the group and moved to the southwest in an effort to get to the creek bed where he would then bend back to the northeast. By doing so, the men hoped to catch their quarry between them in a pincer movement. The fourth member of Echo Team, Alton Higgins, stayed near the old camper sitting below and just to the west of the south cabin in the hopes of catching a glimpse of anything that might outflank the others.

Colyer, with Diaz trailing behind, slowly advanced to a spot where the trail bent gently back to the northwest. He had continued to hear movement in the woods ahead and believed it might be the rock-thrower. After rounding the bend in the trail, Colyer caught sight of his quarry. The figure was large, walked upright, covered in brown hair, and had a cone-shaped head: an ape. Colyer could not see the face of the animal, as it was turned away and looking back down the trail to the west, but he did note that the hair on its head appeared longer than the hair on the body.

The creature was no more than 90 feet distant but was quickly moving away to the south. Colyer also noted what he described as "shadowy movement" immediately behind the ape, as if it was being followed closely by something much smaller. A thousand questions briefly flashed across Colyer's mind, but he had no time to ponder on them and took aim. He quickly fired nine times—emptying his weapon—in the animal's direction.

Diaz rushed to Colyer's side from around the bend in the trail. Bluish white smoke hung in the air for several seconds, obscuring the target, but as the smoke began to clear, there was no sign of the animal. The men hurried to the spot where they assumed the ape had fallen and was lying in the thick vegetation off the trail.

Lawrence heard the shots and recalled: "I was absolutely overcome with adrenaline when I heard those shots and instantly knew that Daryl had encountered a wood ape." Lawrence knew there was a high probability that Colyer had hit his target but realized the animal might not yet be down for the count. At that point, he decided he would rather have his semi-automatic 12-gauge shotgun instead of the scoped rifle he had grabbed initially and ran back to the south cabin to swap weapons before joining Colyer and Diaz.

Meanwhile Colyer and Diaz arrived at the spot where the ape had been moments before, fully expecting to see a body lying on the ground. But there was no body. They did locate several tracks and the tree that seemed to have absorbed the majority of the rifled slugs Colyer had fired. About this time, the two men heard what sounded like the engine of a vehicle starting up from the direction of the property gate, about a quarter mile to the west. Colyer asked Diaz to go investigate the vehicle sounds while he continued to search the area. Moments later, Lawrence arrived at the scene. The men paused only long enough for Colyer to reload (with slugs only this time) before resuming the search.

The two men were joined shortly thereafter by Alton Higgins, and together they combed the woods, frantically searching for a body or blood trail before darkness fell. But their efforts would be fruitless. During this search, Diaz returned carrying a small container of iced tea that he had found near the property gate. This puzzled the team greatly. Emotionally spent, the men returned to the south cabin for the night.

In the morning the team resumed the search. But no body or blood was located. Chagrined, the investigators returned to examine the four tracks they had previously found at the shooting site. The prints measured 16 inches in length. The heel-to-heel measurements were 64, 59, and 52 inches in length. One print showed clear toe impressions. Having gathered what evidence they could, the team left the valley shortly before midday, bitterly disappointed at how close they had come to achieving their goal.

It was not long after Echo Team left the valley that word reached the group leadership that a family member of one of the property owners was very upset with the group. He had parked his vehicle just to the west of the property gate and had begun hiking east along the ATV trail when Colyer had spotted and fired at the ape. The family member had been at least a quarter mile from the shooting location and, due to the manner in which the trail bends, was never visible to Colyer or in the line of fire. The gun shots had badly frightened the man and he quickly turned and ran back to his vehicle. It was his truck the team had heard starting up while searching the woods for a downed ape.

In his haste to leave the area, the family member had damaged his vehicle. The road into the valley is amazingly rough and can damage a vehicle even at slow speeds. It is this part of the incident that has been the source of much misinformation. The organization did offer to pay for the repairs to the family member's truck; however, this was not the result of any sort of litigation or the threat of it, as some have suggested. Everyone in the group felt terrible about the unfortunate situation and the Board of Directors felt that making the offer to repair the vehicle was the right thing to do. The gentleman accepted the money and repaired his vehicle, but the family also decided to part ways with the group. No longer would the NAWAC have access to the cabin compound and the apes that seemed to frequent it so often.

The NAWAC members, though greatly disappointed to lose access to the cabin compound, nevertheless decided to press on.

13
Endurance Continues

In the aftermath of the Echo Incident, the main goal of Operation Endurance was amended. The new directive was to recover any and all trace evidence resulting from the shooting. So on the morning of July 9, Foxtrot Team, comprised of Mark McClurkan, Brad McAndrews, Shannon Mason, and Shannon Graham, arrived in the valley. With the cabin compound now off limits, the group pitched tents about 100 yards west of the property boundary.

The team did not have to wait long for activity to ramp up. During the afternoon they heard cracking sounds, metal impacts, wood-knocks, and rock-clacks, all emanating from the old compound. The team was frustrated at not being able to investigate but at 5:34 p.m., the team observed a "medium-sized rock" land near their vehicles. The rock bounced twice before coming to rest less than 20 yards from the trucks. At this, McAndrews moved west down the road in an effort to observe the rock-thrower. As he did so, McClurkan, Mason, and Graham all heard an animal take "three clear steps" to their south. McAndrews returned in short order and reported that "something was shadowing him." The sound of rocks flying into camp, breaking branches, and movement were heard the rest of the afternoon and evening.

By 9:15 Monday morning, Mason, Graham, and McAndrews—all of whom had work and/or family matters to attend to—left the valley. McClurkan would finish the week alone. Realizing that camp security in black bear country was even more important when camping solo, McClurkan began setting up an early warning bear bell system he had devised.

At 11:31 a.m. McClurkan heard a heavy rock land in the creek to the west of camp and the softer secondary impacts of smaller rocks and gravel that had been scattered when the larger stone had landed. While looking in the direction of the creek, McClurkan spotted "a large brown creature" crossing the creek at a brisk pace. Brush obscured the top and bottom of the figure, but in a window through the vegetation he saw what appeared to be the torso of a large, hair-covered animal. He observed no up and down "bounce" as the animal walked across the creek. McClurkan specifically mentioned the fluidity of its movement in his field notes. He estimated the height of the animal at seven feet and described its hair color as "cinnamon." McClurkan immediately moved to investigate the area where he had seen the figure but found no sign of any kind and returned to camp where, for the rest of the day, he was bedeviled by the sounds of a large animal moving through the brush near the creek—always just out of sight—and metallic impact sounds at the cabin compound.

The next morning, McClurkan was awakened by the sound of "triple wood-knocks" from just north of his location. Later, he documented a "very loud and strange roaring sound" that originated from somewhere on the north mountain. The roar was loud enough that it caused birds to take flight and scatter.

At 9:28 a.m., McClurkan was investigating some odd tree breaks that he and McAndrews had found earlier in the week when he came upon a clearly defined bipedal trackway. The trackway consisted of impressions measuring 18 inches in length and 6 inches, or more, at the ball. The stride measured about 60 inches. McClurkan wrote in his field notes: "These tracks were very fresh as the crimped and bruised grass was still dark and wet where it had been damaged." In all, he found more than 30 individual tracks over a 150-200-foot section of the game trail before coming to an end in the rocks of the dry creek bed.

On the morning of July 12, McClurkan was jarred awake by the sound of what he described as "three echoing moans" that seemed to originate from the slope of the north mountain. He would be plagued with more strange sounds throughout the morning, including the "crunching movement" of an animal through the brush, objects—assumed to be rocks—striking trees, and odd "double and triple bangs" that he described as being akin to the sound of "someone beating on a steel 55-gallon drum." Perhaps the oddest sound McClurkan heard that morning he described as "like popcorn popping... possible bear popping its jaws?" The odd sound was heard again just three minutes later. "The popping has continued and is moving south," McClurkan wrote in his field notes.

Around 11:45 a.m., McClurkan began to feel ill and decided to load his gear and leave the valley. An hour later, he had only one piece of equipment left to retrieve, a trail camera positioned along the creek. After hiking to and removing the camera, he sat down on the bank of the creek to rest. It was then that he noticed a rock across the way that appeared to have a drop of blood on it. It was not far from the southwest corner of the westernmost cabin in the compound, and just to the south of where the Echo Incident had taken place.

NAWAC

Blood drops on rocks located by Mark McClurkan after the Echo Incident.

McClurkan rose to inspect the drop and noticed it had a slight directional splatter. He followed the dry creek bed in the direction the splatter, which seemed to indicate the animal had been moving. About 50 yards further he found a dozen more blood drops on rocks in the creek bed. McClurkan photographed all of the rocks and collected five on which blood had dripped and spattered for future analysis. He continued his search, and though he found no more blood, he did find a series of tracks near an old cistern well southwest of the two-story cabin. While most of the tracks were little more than large, degraded impressions, one was pristine and showed five clear toe impressions. The track measured 16 inches in length. After documenting this new evidence, McClurkan, who was feeling steadily worse with every passing minute, left the valley.

Juliet Team, comprised of Brian Brown, Chris Buntenbah, Daryl Colyer, and Brad McAndrews, rolled into the valley at 6:00 p.m. on Saturday, August 13. Other teams had been in the valley since McClurkan's discovery of the blood-spattered rocks and they too had documented the rock-throws, bangs, knocks, and whoops that had become almost mundane to NAWAC members. One unusual vocalization, detailed by Paul Bowman of India Team,

111

was noted and described as "similar to the Tahoe Scream but with less pitch. Disturbing, like a crazy banshee woman."

Using an ATV Polaris provided by Bowman, Juliet team decided to move base camp up the slope of the north mountain where so many odd sounds seemed to originate. While shuttling equipment up the slope, McAndrews and Colyer were startled by a loud bang on the roof of the ATV. The pair stopped and looked around. There were no overhanging trees from whence a branch or nut could have fallen. The pair could only shake their heads and continue their work.

That night, the men noted a curious "twirling sound," which no one could identify. Brown reported faint "whoop-like" sounds shortly after midnight southwest of camp, in the direction of the old cabin compound. Things quieted down afterward and the team turned in around 1:00 a.m.

At approximately 3:00 a.m., Brown heard movement sounds in the brush line. Armed with his MiNi-14 night-vision unit, he peered cautiously out of his tent and saw a large, vertical "something" move behind a tree. "Whatever it was, it was big and elevated, not low to the ground," he said. Brown told his teammates what he had seen, so the men decided to leave their tents and investigate. Colyer unpacked a second thermal unit that had been donated by Paul Bowman, and he and Brown investigated the area where the vertical figure had been seen. They found nothing. The rest of the night was uneventful.

The team would spend the next two days taking long hikes, call-blasting, and searching for signs. They were rewarded with the find of a "seven-inch, barefoot, humanoid track" in the soft mud of a tributary channel of the main creek and heard multiple wood-knocks, including one that Buntenbah and McAndrews described as "terrifically loud resembling a shotgun blast."

On the morning of the 17th, the investigators awoke to find the right rear tire of the ATV was flat. Knowing they needed the ATV to ferry their equipment back down the mountain, the crew abandoned their plans for the day and trudged down the slope to their vehicles for a trip into town for the supplies needed to repair the tire.

It was 4:45 p.m. before the men, armed with several cans of Fix-a-Flat, were able to begin their trek back into the valley. Nearly an hour later, after cutting saplings and blazing a new trail around a large downed oak tree that blocked the road, Colyer returned to his truck and started the engine. At that moment McAndrews and Brown clearly heard a loud and extremely odd vocalization. Brown frantically signaled Colyer to kill the engine and get out of the truck. Colyer then also heard the strange sound. The men all agreed that the sounds had to have been made by a large animal that, as incredible as it seemed, was attempting to mimic the sawing sounds the men had generated when cutting down the saplings.

"It was a continuous and deep in-and-out wheezing along with a hooting sound," said Brown, who believed the intent was to warn other apes in the area of the presence of the men. McAndrews felt the creature may have simply been exhibiting excitement or attempting to copy the sound of the sawing that had taken place minutes before. Investigation of the immediate area provided no additional information.

The men then resumed their trip back up the North Mountain, where they fixed the flat ATV tire. A thunderstorm blew in a few hours later, which kept the team in camp and prevented any night operations. Juliet Team awoke and broke camp at 8:40 a.m. on Thursday, August 18.

Operation Endurance had come to a close.

14

Operation Persistence

The winter months were spent planning, developing new strategies, and trying to learn from the mistakes made the previous summer. We all agreed that the Holy Grail of Bigfoot research—the securing of a type specimen—was within the reach. With the rosters full, Operation Persistence kicked off in the Spring of 2012.

The NAWAC members were chomping at the bit to get back into the valley. Adding to the excitement was the news that the owner of the north cabin, who was more understanding regarding the circumstances surrounding the Echo Incident, offered to let the NAWAC operate out of his structure. The idea of sheltering inside four walls again and having a roof over one's head does wonders for morale when out in a place as remote as Area X. The group was grateful for this generous offer.

Alpha Team, consisting of Daryl Colyer, Rick Hayes, Alex Diaz, and Chris Buntenbah, arrived in the valley on the afternoon of Sunday, May 5. Things got strange right from the start. While setting up camp, all four men heard a faint but unmistakable "whoop" from approximately 200 yards to their northwest. Colyer and Hayes immediately left to investigate. As they approached the south cabin—the two-story structure out of which the group had operated the summer before—they heard the sound of a large animal crashing through the woods behind the building. The pair rushed around the corner of the cabin but saw only the still-shaking branches of the brush on the opposite side of the creek where something had dived into the woods. "We didn't see it, but by the sound of it, it was a substantial animal," Colyer said.

Diaz, back at the base cabin, had grabbed a pair of binoculars and, although his view was partially obscured by trees, had managed to catch a glimpse of a large, gray, upright figure darting away from the cabin area and into the creek bed. It was clear the creature was retreating from Colyer and Hayes who arrived at the cabin only seconds later. Upon hearing the news, Colyer smiled and said, "Hold on to your butts; it's gonna be a wild week."

Other than a few light wood-knocks and small impact sounds on the cabin roof, the night passed quietly. After breakfast, the men went about establishing two observation posts (OP1 and OP2) in and near the dry creek bed to the south and west of the two-story cabin. The team also deployed an array of Plotwatcher time-lapse cameras along the creek. After the disappointing performance of the motion-activated cameras the summer before, the group had decided to try a different strategy. The Plotwatchers were programmed to take one photo per second during daylight hours. While the group was losing the possibility of capturing a nighttime photo, they hoped that by having cameras continually snap photos during the day, an ape would be photographed as it crossed the creek, something the NAWAC suspected happened multiple times each day.

As the investigators were relaxing after the activities of the day, Colyer happened to glance down the trail that leads to the west cabin. His account reads: "I saw two branches wagging. As I was watching them, a black animal—I'm guessing it would have been about waist high to me—darted across the path like lightning, just a black blur. Rick and I got up to investigate. About 30 feet from the spot where the animal had crossed, another animal broke through the vegetation to my right and ran north toward the slope of the mountain. I never saw it, but it crashed like a bulldozer busting through the brush."

At 9:21 p.m., the two investigators were inside the cabin recording the GPS coordinates of the Plotwatchers on the group laptop when a huge sound startled them. "It was like someone had thrown a Volkswagen," said Hayes. "It was unnerving." The monstrous crash was quickly followed by the sound of bipedal footfalls from the east side of the cabin. Despite being a bit shaken, the pair exited the cabin in the hopes of getting a glimpse of what was causing all of the commotion. Moments later, Hayes and Colyer were startled by a "loud screaming roar and the sound of something furiously banging on metal" to the east of camp.

Paul Bowman arrived in camp at 4:30 p.m. on the 7th. He erected a 24-man Kifaru teepee just southwest of the base cabin near the trail that cut through the woods leading to the south cabin. A few hours later, Ken Helmer arrived; the team was now at full strength.

Between 7:45 and 10:00 p.m., the men sat around a small fire and were treated to the sounds of animal movement in the brush, rock-clacks, wood-knocks, and pungent odors. "A musky zoo animal-like smell drifted into camp periodically," Bowman wrote in his notes. "It would be there for a few minutes and then it would be gone." At 10:15 p.m., Colyer and Helmer observed a pair of reflective eyes about 25-30 feet back in the wood line just west of the cabin. Initially, the men thought they were seeing the eyeshine from one of the camp foxes, but after a few moments they realized the eyes were at least three feet off the ground. "The eyes were greatly reflecting the firelight and were a vivid green," said Colyer. After a few moments, Helmer said, "Let's light it up." Colyer agreed and the men clicked on their headlamps in unison.

"At that precise moment," wrote Colyer, "the animal began to smoothly rise up to a height of what had to be more than seven feet and then moved briskly toward the mountain. Helmer ran to get his spotlight, and I ran after the animal. I saw the eyes turn to look toward me again just as it reached the base of the mountain."

Helmer and the others quickly joined Colyer at the foot of the mountain. As they lit up the slope, a rock soared through the air above them, clipping branches as it went, and landed with a crash behind them. The animal then made a loud, crashing run through the brush higher up the mountain and was gone.

As the team returned to their chairs around the fire, a rock smashed into the east shed. "It sounded like a damn cannon going off," Colyer said. Moments later, the men were jolted again, this time by a raspy scream that originated to their south somewhere near the two-story cabin.

This event, known as the "Bright Eyes Incident" among NAWAC members, brought to light how close these animals could, and would, get to us in order to observe our behavior. While the realization was unsettling, it did solidify in our minds that the innate curiosity of these animals might be the trait we could exploit in order to achieve our goal.

On the morning of Thursday, May 10, while prepping for the day's activities, the team members were stopped in their tracks by a very loud, sustained crashing sound up on the slope to their northwest. "It was like the sound of explosive fireworks with a very long reverberation through the valley," Colyer wrote. "It had to have been a boulder throw."

At about 12:45 p.m., Colyer deployed in OP1 and Hayes began the process of changing out the memory cards on the Plotwatchers lining the creek. Helmer and Bowman left the valley to check in with the NAWAC Chairman, Alton Higgins, by phone. Only an hour after settling into the blind, Colyer was startled by the kind of up-close visual few have experienced. He had been closely monitoring Hayes's progress down the creek to the west, but turned momentarily to check out a small sound behind him. As he was turning back to the west, Colyer saw a large, upright, bipedal animal 50-60 yards down the creek from his position in OP1. Maybe the animal had been trying to keep an eye on Hayes and had not realized Colyer was concealed in OP1?

"The animal," Colyer wrote, "was gray with light-colored feet (possibly the soles) and some sort of light or white coloration in the buttocks area. It moved quickly, fluidly, and smoothly and stepped from a mossy green boulder up on the bank near Plotwatcher 2. The animal quietly disappeared behind the thick wall of vegetation on the south bank. It was only a fleeting glimpse, and within a second the animal was completely gone. It appeared very agile and its movement was natural and easy."

Unfortunately, Plotwatcher 2 did not record the event. We had a limited number of Plotwatchers and they were spread out a bit too far apart as we were trying to cover as long a stretch of the creek as possible. This left gaps in the coverage.

When Bowman and Helmer returned to the valley, the men returned to OP1 so that Colyer could point out exactly where the ape had crossed the creek. Hayes stood in for the animal and attempted to duplicate its creek crossing. At this point, Colyer, realizing the creature he had seen had been much larger than Hayes, became quite rattled. "It just dwarfed him [Hayes]," Colyer said. "It was clearly no less than two feet taller than Rick." Hayes was unable to duplicate the ape's crossing; he could not get from the mossy boulder to the creek bank without leaping. The animal Colyer saw had traversed the distance easily and smoothly.

In camp that night, at 11:10 p.m., a softball-sized rock struck the ground only 10-12 feet behind the spot where Colyer was sitting. Already on edge due to his visual earlier in the day, the proximity of the rock-strike angered him. "I was pissed," he said. Telling his teammates, "They drew first blood," Colyer picked up two rocks and threw them as far up the slope northwest of the cabin as he could. For the next three hours, the men traded rock throws with something up on the mountain slope.

"It was pitch dark," Colyer wrote in his field journal. "I couldn't even find a rock to throw without the use of my headlamp, yet something up on the mountain slope was apparently finding rocks with ease and throwing them with authority toward us in the black of night without the use of lights." The team also noted a "musky odor" that wafted into camp from the east from time to time, an "ooooo" sound emanating from the mountain slope, and some faint chattering sounds from the northwest.

The men, exhausted, turned in at 2:30 a.m. and enjoyed a quiet night after experiencing so much action earlier that day and evening. The activity the rest of the week slowed considerably.

Bravo Team consisted of multiple NAWAC investigators who stayed in the valley for varying amounts of time during the week of May 12-19. Those who were present would have a week to remember.

At approximately 5:45 p.m. on Saturday, May 12, Ken Helmer was concealed in the blind at OP1, while Mark McClurkan, and Ken Stewart were servicing the Plotwatcher cameras. Shortly after his teammates completed their task, Helmer heard a clear "whoop" vocalization an estimated 50 yards to his southwest. Less than 30 minutes later, he saw a dark animal emerge from the woods.

"Down the creek to the west, maybe 100 yards, or so," reads Helmer's account, "I saw a black, upright figure step into the creek from the north bank and look toward me. After a few seconds it disappeared back into the foliage on the bank. Within just a few seconds, the figure re-emerged into the creek and again stared toward me. I could clearly see its arms dangling on the sides and the clear separation between the legs. It was entirely black, face and all. After four to five seconds, it disappeared again into the foliage along the bank. I raised my .30-06 to glass it but it never reappeared."

Upon returning to camp, the men reviewed footage from the GoPro camera Helmer had been wearing. The team was thrilled to find the whoop vocalization had been recorded. But they would be deeply disappointed moments later when they learned the GoPro's 8GB memory card had filled up prior to Helmer's sighting. "Blackie," as the ape has come to be known, had not been captured on video.

On Sunday, Helmer left for home, and Brian Brown, Bob Strain, and Kathy Strain arrived, joining McClurkan and Stewart. About midnight, the team was treated to a high-pitched "wooo" vocalization from somewhere on the mountain slope. More intense activity would soon follow, specifically late Monday afternoon.

At 5:11 p.m., McClurkan and Kathy Strain heard five faint howls in succession. "They were faint but clear," McClurkan said. "They lasted about four to five seconds each." An hour later, the unmistakable crash of a rock on the metal roof of the east cabin was heard. Brown, Stewart, and McClurkan started east down the trail to investigate. A small rock hit a tree near the team soon after they started their hike, but instead of peeling off to investigate, the men continued toward the east cabin where the first impact sound had been heard. The trio of investigators heard movement on the slope; it seemed to be headed west back toward the base cabin. After a momentary pause, as the team continued east, McClurkan caught a flash of something light-colored moving north. The animal crossed the run-off channel behind the east cabin and bolted to the north toward the mountain slope. McClurkan hailed Bob Strain on the radio and asked him to join them. Strain was carrying a big bore .458 rifle; McClurkan thought it would be the weapon of choice should a collection opportunity present itself as the animal had embedded in thick brush.

Shortly after Bob Strain left the base cabin, Kathy Strain, who was on the small east porch watching and listening, spotted movement at the base of the slope just beyond the small shed next to the cabin. "Something peeked out from behind some branches and then let them go," she said. "I got the impression whatever it was realized I saw it and retreated back into the thicker foliage." Strain could not positively identify what she had seen, she knew only that she had "seen branches move and there was something behind them." Upon her husband's return, the Strains investigated the location where Kathy had seen the "peeker" but found nothing.

At 7:45 p.m., the group heard movement to their east that did not match the shuffling sounds they had become used to hearing when the foxes were in the vicinity. As they turned to look, Brown and McClurkan noticed a small black figure standing outside the brush line at the foot of the slope. Seconds later, Kathy Strain stood up and shouted, "Look! There they go!" Brown, McClurkan, and both Strains saw two upright animals bolt up the slope.

"They had to be moving close to 20 miles per hour up that mountain," Brown said. The investigators said the animals had "well-defined muscles and leaned into their stride as they ascended the slope, and splayed their feet—as if for traction—as they climbed." The team was amazed at the speed of the animals; they agreed that no human could ever duplicate their agility and smoothness of movement while ascending the steep slope.

The team continued to monitor the mountain slope and were quickly rewarded for their effort. McClurkan watched a black ape walk across the first plateau in an east to west direction. He had a good, albeit brief, look at the animal from the waist up. Moments later, Brown called out that he had seen "some kind of movement higher up and to the west" from the spot where the two apes had been seen scaling the mountain. While scanning the area pointed out by Brown, McClurkan spotted an ape that briefly side-stepped out from behind a tree near the top of the slope. McClurkan saw the animal clearly for about two seconds before it stepped out of sight. "I saw the light showing under its arms, between its legs, and above its head," he said. "In the time it took me to raise my scope from about the level of its abdomen to the vitals, it disappeared. There was no time for a shot."

On Thursday the 17th, Brown and McClurkan heard what they described as a "deep, guttural growling" from an area behind the east shed. The growl was quickly followed by movement in the brush when the pair investigated. Later, the team members heard a strange chattering from the northwest. Kathy Strange described it as being similar to "a woman talking."

At 12:40 a.m., a loud "thunk" noise, similar to "wood-on-rock" was heard from somewhere near the south cabin. Moments after that, "two sharp snapping sounds" were heard from the same vicinity. When Brown, Stewart, and McClurkan went to investigate, they were greeted by the now familiar "horse smell" and detected movement ahead. What they saw was indeterminate, only a shadow, but something had moved near the cabin. Cautiously, the men advanced until they were startled by a loud noise. Brown said, "It sounded like something bumping up against the green shed." Quickly thereafter, the investigators detected more movement down in the creek to their east and back to their northwest near the trail. It was now obvious there were two animals in their immediate vicinity. The team then heard what Stewart described as "an exasperated, deep-toned, groaning exhale." The men lit up the area with white lights but saw nothing.

"We should see it RIGHT THERE!" said a frustrated Brown. McClurkan urged his two partners to advance slowly. They reached the point where the forest floor gives way to the creek bed below, stopped, and then flooded the area with white light. They were all positive that the two animals were extremely close, hiding, yet standing their ground. Realizing they were in a dangerous position, caught between two apes they could not see, they decided to carefully retreat and head back to the base cabin.

But things did not settle down once back at base camp. Between 1:30 and 2:10 a.m., the team experienced multiple rock-throws, the sound of bipedal running footfalls, and growling from the mountain slope. At 2:20 a.m., Brown looked back to the east and observed an intimidating figure just outside the wood line at the base of the mountain. "I saw a dark mass, about seven feet tall at the extreme end of the clearing," said Brown. "As I watched, I detected a very slow swaying motion." After alerting the others, McClurkan and Stewart illuminated the area with white light and also observed the figure. Immediately there was a great commotion up on the slope. Initially, they ignored the racket and stayed focused on their target but eventually glanced toward the noise. When they turned back, the original figure was gone.

The night still had a few more surprises in store. About 2:33 a.m., Brown remotely turned on his truck lights in order to scan the area behind the cabin.

In reaction, a large animal—no more than 15 feet in front of his truck and 30 feet from the group—tore off through the woods to the east. "For a while," Brown said, "there was so much movement and commotion on the slope of the mountain the team did not even bother to record the incidents in their journals. "We couldn't keep up." At 3:40 a.m., Brown moved away from the fire to relieve himself and encountered "something big" moving within the tree line to the west. "It was large, and uncomfortably close," Brown said. Exhausted, and "having had enough," the investigators retreated inside the cabin and went to bed.

Rest would not be in the cards, however. At 4:45 a.m., Brown heard movement outside the east wall through the small window above his bunk. He reported small rock or nut impacts, footfalls, and "small metallic sounds." What really got under Brown's skin, however, was when he heard and felt what he perceived as something leaning on or pushing against the northeast corner of the cabin. Brown wrote, "The entire structure audibly shifted. At that time, I climbed out of my bunk so as not to be next to the 'Boggy Creek' window and finally got some sleep in the recliner in the middle of the room." McClurkan and Stewart, who had been snoring loudly when the incident occurred, both woke up after feeling the shift of the structure.

At 7:00 p.m. the team gathered to discuss their plan for the coming night. Based on the level of activity and the number of visuals over the previous 48 hours, the possibility of a collection opportunity seemed very real. But other than the sounds of a bobcat caterwauling and a flash of "something black" moving on the slope, the night was quiet and no such opportunity presented itself.

The team was awakened on the morning of the 18th by multiple rock-impacts on the cabin and surrounding sheds. At 10:00 a.m., Bob Strain caught sight of a large figure walking up on the mountain slope. "It was reddish brown; I caught sight of it as it walked through a sunny spot. It was traveling to the west," he said.

That evening at 7:10 p.m., McClurkan was in the blind at OP1 when he began hearing "short-range knocking or popping sounds." He could not tell how these sounds were being produced but did say, "They sounded quite a bit like the sound produced when you pop your tongue." At 7:15 p.m., he heard the sound of a large animal cross the creek approximately 30 yards behind him. He alerted Brown that, by the sound of it, it was headed in the direction of the east cabin. A short while later, the odd clicking/popping noises began again. The sounds were coming from the south and then from the west, as if in response to one another. After listening to these sounds for some time, McClurkan decided to respond himself and "clicked" his tongue loudly in an effort to mimic them. Immediately, the animal to the south growled fiercely. "It was deep, guttural, menacing...a warning," McClurkan wrote in his journal. Deciding that being sandwiched between two large animals—at least one of which that was now clearly agitated—was not where he wanted to be, McClurkan radioed Brown and asked to be escorted from OP1.

As Brown and Stewart headed toward OP1, McClurkan heard the animal beginning to move through the brush toward his position, prompting him to radio his request for extraction again. Within seconds, things took a frightening turn.

"I heard the animal to my south picking up the pace as it moved closer," said McClurkan. "I heard the loud thumps of heavy feet start slowly and rapidly increase in tempo as the animal began running. It was charging from an awkward angle, approximately my seven or eight o'clock position. I attempted to stand up and turn to face the animal and when it got to within 40-50 feet, I could make out a figure that was upright, black, and on two legs charging toward my position. I could hear its feet hitting the ground and the sound of vegetation being pushed aside. It was moving at least as quickly as a deer. Due to the position of my chair inside the blind, I could not get my body completely turned to face the animal, so I just twisted my upper body as best I could and fired. The ape immediately turned away, it actually slipped because all of its momentum had been coming in my direction. I chambered another round and kicked the chair and blind panels down so I could get fully turned. I fired another shot in the direction of the animal but am certain the round hit a tree."

Brown and Stewart were in a near panic after hearing the two rifle shots and were now in a dead sprint in an effort to reach McClurkan. They met up with him in the woods on the north bank of the creek, all with frayed nerves. The rest of the night was quiet.

After McClurkan's terrifying experience, Field Operations Coordinator Daryl Colyer placed a hold on the next team that had been scheduled to enter the valley. Instead, a non-member named Robert* was given the opportunity to collect a specimen. Robert owned some specialized equipment including two high-powered rifles with thermal scopes and several thermal cameras that could be deployed for passive monitoring. These thermal cameras could be used in tandem with remote monitors so that a person could keep watch over an area from the relative safety of the base cabin. While the NAWAC typically did not allow solo operations in Area X, Colyer elected to allow Robert to do so based on some thermal footage he had obtained while working alone in the Sabine National Forest of East Texas.

Robert arrived at the compound on the night of May 19 and was briefed by Bravo Team members before they exited the valley early Sunday morning. For the next two days Robert attempted to establish contact with the apes of the valley without success. A few days later Robert would express his skepticism to the idea of apes in the valley. When Jerry Hestand, Mark Porter, Brad McAndrews, Travis Lawrence, and Daryl Colyer arrived the evening of the 25th, Robert wasted little time in approaching Colyer and throwing cold water on the enterprise. "There are simply no apes in the area," he said confidently, "and what you guys are reporting as rocks hitting the cabin are most likely nuts dropped by birds and flying squirrels."

Colyer politely, but firmly, replied: "I know the apes throw rocks because we had a two-hour rock war with the bastards." He then reminded Robert that there had also been multiple clear daylight sightings by NAWAC members over the last ten days.

Around 9:45 that evening, a clearly disturbed Robert walked into camp and rather sheepishly informed Colyer he had seen an ape walk across the trail between the south cabin and the east cabin a few minutes prior. The animal had walked into the middle woods of the compound. The team immediately set out to intercept it and encountered an animal concealed in brush. It gave a "loud, deep, rumbling, threatening growl" but the team could not flush it out and eventually returned to the base cabin.

Just before 6:00 a.m. the next morning, the entire cabin was rocked by "a tremendously loud and enormous impact on the side of the structure." The men, all awakened by the impact, jumped out of bed and rushed outside with weapons drawn. They saw nothing. Lawrence was sure the impact had occurred on the west wall of the cabin. He said, "My shotgun had been leaning against that wall and came crashing to the ground after the impact." Leeper wrote in his field notes, "To me it was like it had just put its shoulder down and rammed the wall of the cabin."

The men, adrenaline pumping, were now wide awake, so they conducted an inspection of the west wall of the cabin. They found no damage; however, Robert found a new and good-sized rock up on the cabin roof. Appreciating the irony, Colyer said, "That was some big-ass flying squirrel wasn't it?" Chastened, Robert said nothing. He would leave at the end of the week with the rest of the team, having a new outlook on Area X.

Delta Team members Tod Pinkerton, Rick Hayes, and Dale Ryan arrived at base camp at 4:00 p.m. on Saturday, May 26. Around 3:00 a.m. the next morning, Hestand and Porter heard the sound of footfalls and the thud of rocks landing on the ground nearby. At 5:40 a.m., the entire team was awakened by three successive rock-impacts on the cabin roof. The barrage of rocks continued, off and on, until shortly after 10:00 a.m. Later, when McAndrews decided to sweep the roof, he located four stones. "Rocks can't just jump up on a roof," he said.

The next few days were quiet, at least by Area X standards, with only the occasional wood-knock heard. By the morning of the 30th, only Pinkerton and Ryan remained in camp to welcome the incoming Shannon Graham and Shannon Mason; Pinkerton and Ryan would exit the next day, leaving the two women alone in the valley for several hours. This was the first time that female investigators were deployed alone in Area X. How might the resident apes react?

"Us girls are here alone now," Graham yelled loudly up the mountain slope. "You guys should come out and play, party, and raise hell!" Moments later Graham's "invitation" was answered by a single, but very clear, wood-knock from the northwest. The two investigators would record multiple wood-knocks on a TASCAM audio recorder over the next few hours.

NAWAC Chairman, Alton Higgins and Jason Hill arrived in camp at 11:30 that night. Later that morning the team found three of the Plotwatcher cameras along the creek had been "twisted around." They showed nothing but blurred images of the area immediately around the tree on which they were deployed. It would appear that whatever moved them did so from behind. This is not as unlikely as it sounds; Siberian tigers have been known to do the exact same thing. The investigators repositioned the cameras. Graham and Mason departed for home soon after.

At 5:00 p.m., Higgins set about rigging a string trap across a game trail behind the large teepee Paul Bowman had set up earlier in the operation. Higgins tied one end of the industrial black thread to a tree at a height of six feet. He stretched the thread across the game trail and wrapped the other end of the thread around another tree.

"The idea is to see if something tall is walking that trail," Higgins explained. "The thread is all but invisible to the naked eye and anything that walks through it will cause it to unwind on the non-tied end. The thread will then stretch out in the direction the animal was traveling." The use of such thread or string traps would figure heavily in future events.

Over the next four days, the team documented dozens of anomalous sounds, including wood-knocks, rock-strikes, "shrill piercing screams," footfalls, what Higgins described as a "nearly continuous rumbling, reminiscent of something shaking sheet metal or tin roofing," and "odd, moaning, melodic vocalizations." Additional string traps were set and one in particular, which had been strung between the northeast corner of the cabin and a large hickory tree, had been disturbed on two different nights.

After retiring to the teepee on the night of the 8th, Higgins wrote the following passage in his field journal: "Last night was very active…didn't have my journal, so I wasn't able to keep track. The apes are here; I've no doubts, and they may even be near as I write. Just within the last hour I've twice flushed a large animal standing just off the path near the base cabin, and rocks have been thrown. It's important to note that I can hear really well from inside the teepee; it's the same as sleeping outside. I heard the sounds of something coming down the mountain, knocking rocks loose in a manner as if to say, 'I don't care who hears me coming.' I heard the disturbance created by huge-sounding creatures moving noisily through thick vegetation, large branches moving and sticks breaking, reminding me of what elephants might sound like if forced to walk through this tangled jungle that surrounds us. I heard what sounded like boulders being tossed in the rocky stream beds south of the other cabins. I heard tremendous 'BANGS' as rocks struck metal roofs and/or sides of at least four different buildings within the compound. These events occurred throughout the night. At 0650, the most disconcerting event of all took place. I heard something walking slowly—it must have been on a path or the road—at a pace of about one step per second…a lumbering pace. I wasn't hearing it walk as a result of hearing leaves crunch or twigs snapping; I heard the impact of great weight on the ground, and more than that, I could FEEL the impact as I lay in my cot. It must have been close. A mystery: Why aren't the apes always stealthy?"

At 9:00 a.m. on the morning of June 9, Higgins and Hill retrieved the TASCAM audio recorder from its position on a plateau up on the mountain slope and then commenced a patrol of the cabin compound. Diaz, who remained on the front porch, spotted a large, light-colored creature paralleling Hill and Higgins, who were walking past the south cabin. The two seemed oblivious to their "escort." An hour and a half later, a "drive" through the middle woods in an attempt to flush anything hiding bore no fruit.

Phil Burrows and I arrived at the compound at 4:30 p.m. on Saturday, June 16, joining Travis Lawrence, the cabin owner, his son, and a friend of the son. Within the first few hours we were treated to the sounds of rocks banging off of various metal-roofed structures, a large boulder rolling down the mountain slope, and an odd yell off to the northwest.

Between 11:00 p.m. and 3:00 a.m., the group experienced an incredible rock-throwing event. Time and again, with no more than a ten-minute gap between events, we heard rocks strike the base cabin, the east shed, and other structures in the compound. These rock-throws were often quickly followed by the sound of something moving up on the slope of the mountain. It was really quite remarkable. I sat there in wonder, staring up the mountain, knowing I was within yards of a completely undocumented species.

Things slowed down around 3:00 a.m., and the team turned in for the night. Then at 4:15 a.m., the cabin roof was pounded by "a large rock, or other heavy object." From that moment until 6:00 a.m., the barrage of rocks continued. Despite the impacts occurring every 10-15 minutes, I finally managed to drift off to sleep. At 6:15 a.m., a rock struck the east shed with more force than usual. The impact woke me, and I stepped outside in the hopes of spotting movement in the early light. I was soon joined by the cabin owner. The rock-throwing ceased completely once we were outside.

The cabin owner and the two boys left for home at 11:30 a.m. Shortly after, Burrows and I deployed a video camera on the lower slope of the mountain. The camera was mounted on a camouflaged tri-pod and embedded in brush. I focused it on a window in the vegetation that provided a view of the plateau where multiple visuals had occurred. I felt that a new approach was in order since the trail cameras had been proven unreliable and the apes seemed to avoid the Plotwatchers with the skill of a cat burglar. I thought that focusing on an area a good distance from the actual camera location might prove effective. I repeated this strategy in multiple locations over my week in the valley but had no luck. Burrows and I left the valley on the 21st.

Travis Lawrence returned to the compound on July 3. He would spend two nights alone before being joined by any other NAWAC team members. His most interesting experience took place just before midnight on July 4. He was shaken from a light doze at 11:47 p.m. by the sound of voices outside at the front of the cabin. He got out of bed and walked out onto the front porch fully expecting to greet early arriving fellow team members. But no one was there. Lawrence was understandably a bit confused and disturbed. Realizing there would be no sleep for a good while, he set up an observation post in the east bedroom. Fifteen minutes later a rock struck the roof of the cabin. "It was the first such rock-throw of the day," he said. "I couldn't help but wonder if the apes knew I was sitting up looking for them."

By 4:30 p.m. the next afternoon, Lawrence had been joined by fellow NAWAC investigator Steven Matthews* and his son, Greg*. At 9:43 p.m., the team heard a series of strange, high-pitched, screaming howls at least a half mile away. There was one rock-strike on metal to their southwest at 11:28 p.m., but the rest of the night was quiet.

Between 6:50 and 7:06 p.m. on Friday, July 6, the men heard multiple wood-knocks, all from an area southwest of their position. At 7:10 p.m., a steady rain began to fall. Several minutes after the rain began, the investigators heard a vocalization to their west. "It was odd and loud," Lawrence said. "It sounded like the horn on a steamboat."

At 8:15 p.m., Matthews and son walked to the back of the cabin to have a look up the slope. They heard distinct movement and saw a tree shaking about 15 yards up the hill. Matthews called for Lawrence to join them. Unlike others who had retreated in similar circumstances, Lawrence unholstered his .45 sidearm and marched up the slope directly at the tree Matthews had seen shaking moments before. (Lawrence—who had been badly shaken during his first stay in Area X the year before—had learned to channel his fear into action.) Immediately, Matthews heard the sound of a large animal—only 20 yards away—crashing through the woods. The men attempted pursuit but quickly gave up and returned to the cabin.

At 8:35 p.m., all three men heard a loud snapping sound to their east. Thinking it might be an eventful night, Lawrence set up the Marantz audio recorder behind the cabin. An hour later, Steven Matthews went to bed, leaving Lawrence and Greg on the porch. At 11:45 p.m., Greg told Lawrence, "I keep hearing movement to the west. It sounds like a big animal brushing up against a wet bush." While the two were investigating west of the cabin, a loud crash occurred on the mountain slope. The pair immediately raced to the spot and arrived in time to hear another loud crashing noise from behind them somewhere to the south.

Lawrence went to the cabin and woke Steven Matthews. The trio walked behind the cabin and began scanning the slope. Greg had a handheld thermal imager, Steven Matthews had a powerful flashlight, and Lawrence carried his Remington 870 12-gauge shotgun, which was loaded with slugs. As Steven Matthews scanned the slope with the flashlight, Lawrence caught sight of eye-shine. "The eyes were very large and oval-shaped," he said. "I could see them very clearly, and they seemed to be about the size of tennis balls. I saw them look to the southwest and then sweep back to the southeast, then they were gone." Lawrence instructed Matthews to shine the flashlight on the exact spot where he had seen the eyeshine. Again, the eyes briefly flashed, and Lawrence fired three shots in quick succession.

After the third shot, Steven Matthews saw a figure rise up. "I clearly saw the head and shoulders of an ape; it was huge, with little or no neck and was black or dark brown." Matthews fired three shots from his .40 caliber pistol and was sure he had hit the ape. Lawrence was not as convinced and decided to have the team sit tight for 30 minutes before heading up the slope to look for blood. It is not an uncommon thing for hunters to do.

Just before midnight, Greg Matthews saw movement near the spot where the eyeshine had been spotted. Using the thermal imager, Matthews saw the "shoulder of an ape." Lawrence yelled up the mountain, "I'm coming for you!" and started up the mountain. (Unfortunately, our thermal imagers did not have video capability.)

Matthews called him back when he spotted the eyeshine again. He instructed Lawrence to fire directly at the spot being illuminated by his flashlight. "That ape is right there!" he told Lawrence. When Lawrence fired, the brush erupted with violent movement. "I saw a huge gray arm; the right arm of an ape," Lawrence said. "It was probably a foot in diameter."

Steven Matthews was able to see the upper half of the ape. "I could see its right bicep, the shoulder, and part of the head. It was large and gray," he said. The men heard a menacing growl come from the slope moments after the shot. Lawrence, thinking they might spot the eyeshine again, sprinted around to the front of the cabin to swap the shotgun for his rifle.

At two minutes past midnight, Greg Matthews saw a large animal moving on the slope through the thermal unit. The men took turns scanning with the thermal and located a "large circular splatter mark on the ground, about two feet in diameter, that glowed white-hot on the thermal." There were small white drops on the surrounding leaves and another splatter mark about ten yards higher up the slope.

At 12:13 a.m., despite still hearing occasional movement on the slope, the men decided to head up the mountain in an effort to confirm they were seeing blood through the thermal imager.

"Just as we reached the base of the mountain," recalls Lawrence, "we all heard an enormous crashing sound, heavy movement, and sticks breaking. It was clearly the sound of a huge animal moving quickly straight for us. John centered the flashlight on the trees due north of us, and I aimed my rifle in the illuminated area and waited for the animal to emerge. We could all see the foliage moving and shaking as the animal charged. The ape got about 15 yards from us and then turned and ran back up the mountain. As it did, Greg saw the head and both arms of the ape through the thermal. He described a 'huge pointy head and enormous shoulders and arms.'"

The men had been bluff-charged, which is a common intimidation tactic employed by the known great apes. Lawrence, shaking with adrenaline, shouted profanely up the mountainside at the escaping animal. Immediately, as if in response, the three men heard a threatening, low-pitched growl from their northwest. After several minutes of hearing nothing but the rain, the men walked up the mountain to look for the splatter marks. While they did detect the classic "horse smell," the rain had washed away the splatter marks. Since the rain had intensified, the men retreated to the cabin where they passed the rest of the stormy night.

Once Lawrence got word to the group, a recovery team was dispatched immediately. Bill Coffman, Phil Burrows, Mark McClurkan, Ken Stewart, and Alex Diaz arrived on site at 6:30 p.m. the next day and immediately began to scour the slope in an effort to find trace evidence, such as blood or hair, that would prove conclusively that Lawrence and the Matthews' had encountered an ape the night before. McClurkan, a longtime tracker, was able to account for the first three shots Lawrence fired and the three rounds sent by Steven Matthews when the men had targeted "Ironman," as Lawrence had nicknamed the ape due to its unusually bright eyeshine. But the group could not account for Lawrence's last shot of the night, the shot that was immediately followed by the flash of a huge gray ape's arm.

It was obvious to all that at least two large animals had been all over the slope the night before. McClurkan found two areas where it appeared a big animal had lain down. In one, he found what appeared to be finger depressions in the mud. The men searched the slope until dark and then returned to the cabin. The men would search most of the next day as well, but were unable to find any additional evidence that Lawrence had wounded an ape.

On Sunday, July 15, I, along with Daryl Colyer, Ken Helmer, and his friend Tom Ireland* joined Lawrence at the compound at 1:00 p.m. We had been there less than 30 minutes when Colyer and I, walking to the back of the cabin to stare up at the thick vegetation, heard a rock slam the cabin roof, giving us both quite a start. We turned toward the cabin in time to see the rock roll off the roof; it was the size of a softball. "Well, I see they're still here," I said and turned to look back up the mountain. Colyer replied only with, "Yep."

We were all in bed by midnight, but sleep would be hard to come by. At 12:45 a.m., we all heard a strange "bloop" noise. I have absolutely no idea what it could have been. Only 90 minutes later, Colyer, who had placed his cot in the kitchen on the north side of the cabin, heard the odd popping sound described multiple times by other investigators. "It was like a combination between a popping sound and a water drip," he said. At 2:55 a.m., Colyer was jolted out of a light sleep by the sound of footfalls and a very ape-like "huu-huu" sound emanating from behind the cabin. Seconds afterward, the cabin roof was slammed with a rock. The stone bounced and tumbled loudly down and off the corrugated metal roof and landed on the ground with a thud behind Colyer's location. It turned out this would just be the start of the rock party. Between 3:00 and 4:00 a.m., the cabin was hit by rocks repeatedly. "Frustrating," I wrote in my field journal. "There just isn't anything we can do about it."

The next evening, Colyer suggested constructing an overwatch structure out of black plastic trash bags in the bed of Helmer's truck. Colyer had recently learned that thermal units could "see" through this type of black plastic. While we were a bit skeptical, we went ahead and created a makeshift tent using tree branches and a shovel as supports and then duct-taping the sheets of black plastic to them. Once finished, we tested the idea. Looking out of the structure with the thermal, we could see shockingly well. We were all excited about the possibilities this created. We hoped the apes would consider the structure just another tent and conduct their normal nighttime activities, not realizing they could be seen by two team members sitting inside the plastic structure with the thermal unit.

Lawrence and Helmer entered the overwatch shortly before midnight. Everyone else retired to the cabin. At 12:47 a.m., the pair heard two loud crashing sounds similar in volume to the "boom" of dynamite. They also heard wood-knocks on multiple occasions over the next several hours. At 3:05 a.m., they heard "chatter" to their east. They were all but certain they would see an ape before the night was over. But once again the target species proved elusive and, overcome with exhaustion, Helmer and Lawrence left plastic structure and entered the cabin at 4:50 a.m. to get some sleep. A mere 20 minutes after they entered the cabin, a rock slammed the roof of the structure. Lawrence and Helmer could only shrug their shoulders.

By 2:00 a.m. on the morning of the 19th, we were all trying to get some sleep. Colyer and Lawrence were in the bunk beds against the east wall. I was in a cot in the main middle room not far from the front door. Helmer was in the east bedroom, his bed directly against the south wall of the cabin and under a screened, but glassless, window. What happened next is one of the strangest experiences of my life.

"At 3:24 a.m.," I wrote in my field notes, "Colyer and I were awakened by the sound of Helmer talking softly in the east bedroom. He groaned and then said, 'Mike... Daryl... push... help.' Immediately after he finished speaking, I heard the sound of a voice (not Helmer's) speaking gibberish of some kind. The closest thing I could compare it to is the 'Samurai chatter' some have reported in association with these animals. I heard this chatter for just a second, then all hell broke loose. There were impact sounds on the south wall outside the east bedroom and the noise Helmer made when kicking the wall and jumping out of bed. He yelled, 'What the fuck?' When asked what happened, he said 'I was asleep on my left side having a dream... My dream was interrupted by something pushing me away from the side (south) window. I was pushed on my buttocks/hip about a foot or two. I was immediately fully awake and started trying to move away. I then felt a slight pressing down and a pull toward the window on my right hip. I called for help and couldn't figure out why you guys wouldn't wake up. Another moment went by and I tried to get away from whatever had its hand on me. I'm not sure the thump heard was me kicking the wall or something else.' After we were all awake and had discussed what had just happened, Helmer went to the bathroom and saw large 'silver dollar-sized' green eyeshine outside the west facing window. He watched as the eyes turned away to his left and disappeared."

About 10:00 the next morning, while going over what had happened the night before, we decided to take a look at the south facing window of the east bedroom where Helmer had slept; this was the screened window with no glass. Upon investigation, we found the bottom left corner of the screen had been pushed in and was in a bent position, creating quite a large opening. Stuck to the screen and the wood of the sill were several very long white or gray hairs. We collected the hairs and bagged them for possible analysis in the future.

These hairs were later sent to Dr. Brian Sykes when he was doing his big DNA study of possible Bigfoot hairs several years back. His findings were featured in a television documentary, but our hairs were not featured on the show. When we inquired as to what his findings were regarding the hairs we had sent him, we were told they were never examined and their whereabouts were now unknown.

That night, we decided to try an experiment where we feigned sleep.

Helmer played snoring sounds on his game caller, while we sat at the ready inside. But hearing nothing all night, we had all drifted off into a light doze by 6:00 a.m. Suddenly we were jarred awake by a rock impacting the roof and five or six "odd pops." A little more than a half hour later, we heard something substantial moving behind the cabin.

Colyer burst from his seat and out the door; I followed, shotgun in hand. We saw nothing. Since we were already up, we decided to walk the compound loop, each traveling in different directions, in the hopes of spotting something. I was approaching the east cabin when I heard a large animal moving through the woods south of the small tributary. Shortly thereafter, I heard a second animal on the move.

"Moments after hearing the movement," I wrote in my field notes, "I heard a second animal moving loudly south of the tributary from the west. It seemed to be on a beeline for the spot where I had just heard the first animal moving about. There was an opening in the brush directly to my south, no more than 15 yards away. The second animal seemed to be heading right for it. I knelt, raised my weapon, and prepared to make the shot. I thought, 'My God, it's going to be me; it's going to be me.' Just as the animal seemed about to step into the opening it stopped. It just stopped; I don't know why. I knelt there, weapon ready, for more than ten minutes. I crossed the creek to search the woods, as I had not heard the animal retreat. I found nothing."

That afternoon, around 3:10, we all heard a tremendously loud "dynamite-like sound," like those we had heard before but this one seemed much closer. Investigation yielded nothing. At 6:55 p.m., we decided to try an experiment at the creek. I would act as bait—walking up and down the creek, fiddling with the Plotwatchers, singing, and whistling in an attempt to draw attention to myself—while Colyer and Helmer posted up in a concealed position on the heavily wooded bank. As I entered the woods, I heard a low, deep growl and saw something moving to my right (west), the exact location of the Echo Incident. I did not have time to ponder this as I saw a tall silhouette moving at an angle away from me. It was backlit by the sunlight making it impossible for me to distinguish any details. I radioed my team members who came to my location immediately. Knowing an ape was in the vicinity, we decided to proceed with the original plan but never saw anything so we headed back to camp.

After three months, Operation Persistence was drawing to a close. On the morning of the 20th, Helmer packed up and headed home. In the afternoon Colyer and I began the final clean-up of the cabin and the immediate vicinity.

After removing the memory cards from the Plotwatchers, we found a large, barefoot track while searching an area to the west for a Reconyx camera that had never been recovered from Operation Forest Vigil. Impressed in thin soil atop a very hard substrate, the track measured 14 inches long, six inches wide at the ball, and four inches wide at the heel. We photographed the track (the soil was too thin to make a cast) and went back to the cabin, where we proceeded to burn the camp trash.

At 9:49 p.m., we were sitting in our chairs in front of the cabin when we heard a sound the likes of which I had never heard in the woods before. It was an incredibly loud explosion-like sound and came from the creek area to our south. It seemed impossible that a noise that loud could occur naturally. Believing it was some sort of intimidation display, Colyer fired his .45-70 toward the creek. "We can make loud noises, too," he said. I just nodded and muttered "Yes, we can."

The truth of the matter is that this incident rattled our cages pretty good. We were exhausted, dejected, and now more than a bit unnerved. We contemplated leaving that night but felt it best to stay and get some sleep before making the long drive back to Texas. We retired inside the cabin at 11:00 p.m. wanting nothing more than a quiet night.

That was not to be as the early morning hours of Saturday, July 21, proved to be the most unnerving of the week. Between midnight and 5:30 a.m., we were awakened time and again by all manner of noises. We heard what sounded like bipedal running, heavy movement on the mountain slope and to the west, and small rocks striking the roof. It seemed to us that multiple animals were circling the cabin during the night. "It's like they're having a damn track meet out there," I remember saying to Colyer. The most disturbing incident was when we heard something run up to the cabin, slap the wall, and then run off. This was followed by the sound of huffs and growls. "The impact was not as strong as some that have been reported by previous teams," Colyer wrote. "This sound seemed to have a 'meatier' sound, as if the wall had been popped with an open hand."

We left for home early the next afternoon. We were exhausted and dejected; we had not achieved our goal. Operation Persistence was over. The valley once again belonged to the apes alone.

15
Operation Relentless

Operation Relentlesss began on April 26, 2013, with the arrival of Daryl Colyer, Brad McAndrews, Travis Lawrence, Tod Pinkerton, Paul Bowman, Bob Strain, and his friend, Bob Martone*. Shortly after arriving, the team heard what they described as "a strange vocalization." The men looked at each other and smiled. "Here we go," said Lawrence.

The men quickly began construction of an overwatch tent, but instead of building it in the bed of a truck as had been done the summer before, the men erected a canopy—the kind people visiting the beach might use, consisting of four poles and a cover—as the skeleton of the structure. The canopy provided the roof, and the black plastic was stretched and secured to the four poles that formed the frame.

NAWAC

The overwatch tent.

The first few days were not without incident. The team heard whoops, large crashes, "screams and high-pitched howls," "screechy vocalizations," and a "whooo-ahhh" call, but they saw nothing.

The team was up and about early on Wednesday, May 1. The first oddity of the day was a clear wood-knock to the southwest at 10:45 a.m. Several unexplained sounds followed over the next two hours, but the strangest was one reported by Bowman and McAndrews. "It was a sound similar to that of a golf ball bouncing on hard concrete," said Bowman. "It was freaking strange and within 100 yards." Just before 2:00 p.m., the men heard a very "gorilla-like" barking sound to the northwest. Roughly 25 minutes later, the mean heard a sound similar to that of a thin limb being whipped through the air just to the west of the cabin. This was followed by the sound of a large animal moving through the brush and still more "whipping" noises. At 2:59 p.m., the team heard a loud impact sound from the vicinity of the west cabin. Colyer and Strain geared up and went to investigate.

The men approached the west cabin carefully. Colyer crept up to the area behind the cabin to post up while Strain remained on the road. Colyer settled into a recessed area between the cabin and the creek, and after a few minutes he saw two figures moving to the west through windows in the vegetation. As he watched, the animals changed course and turned south, likely to avoid Strain on the road. The creatures splashed across the creek and continued south.

"The animals were close to one another," said Colyer, "one behind the other. The first appeared to be vertical in posture and the second appeared more horizontal. I caught a brief glimpse of leg movement as the animals turned to cross the creek. Both animals were charcoal in color: lighter than black but darker than a standard gray."

Colyer attempted to follow the animals but had no luck. Meanwhile, Bowman joined Strain on the road. As he looked to the south, Bowman spotted what he described as "an upright figure behind the west cabin walking toward us." The figure turned and walked to the northwest until it was obscured by the RV near the west cabin.

"It was reddish-brown around its head, which had a pointed crest," Bowman wrote in his field notes. "The arms were a mottled brownish color. The chest and abdominal area were both of a single-tone dark color. The face was dark, too." The arms swung freely as the figure walked away. Initially, Bowman wondered if it might have been Coyler. But, he wrote, "The figure was colored differently than what I knew Daryl was wearing and it didn't walk like him." Once the men reassembled, it became obvious to all that Bowman had not seen Colyer.

By 1:45 p.m. on Saturday, May 4, the full complement of Bravo Team members had arrived in the valley. The team comprised Brian Brown, Brad McAndrews, Bob Strain, Jeff Eltringham, and Ken Helmer. In addition, Jerry Hestand, Mark Porter, and Daryl Colyer were present for part of the week.

While on a hike to the top of the North Mountain the next day, Brown reported, "two deep, but quiet huffing/hooting sounds" upon reaching the summit. Brown and his team did not believe they were supposed to hear the sounds and wrote, "It was as if the animals producing the sounds were attempting to discreetly alert each other." Back at base camp, the investigators who did not go on the hike reported hearing a squeaking sound from the mountain slope behind the cabin. Later they learned that Brown had placed four squeaky rubber balls on the slope and wondered if an animal had located them.

The next afternoon, Brown and Eltringham hiked up the slope behind the base cabin to check on the four squeezable squeaky balls that Brown had left behind earlier. They quickly located two of the rubber balls; they were exactly where they had been placed. But two of the balls were no longer in the spots where Brown had left them. The search for the balls was brought to an abrupt stop when one of the men took a step and was startled to hear a loud squeaking noise. The ball was not visible, so the pair began clearing leaf litter and forest debris and they located it in short order. Brown had originally placed the ball in a "divot" on top of a large deadfall tree 24 feet away. "By all appearances, it had been deliberately concealed," Brown commented. There were no tooth marks, blemishes, or scratches on the ball. The fourth ball, which Brown had placed on a large square rock nearby, was never found.

On the afternoon of the 10th, Brown and Eltringham were hiking to the swimming hole a half-mile east of camp when they discovered a large, flat boulder with a number of crushed hickory nuts on top of it. Adjacent to the hickory nut residue and shells was a smaller brick-shaped rock that appeared to have been used to crush the nuts. The men were aware that chimpanzees had been documented using rocks to crack open nuts and felt strongly that was exactly what had been taking place at this location. This would be the first of several "nut-crushing stations," as they would be dubbed by the group, to be discovered in the valley.

The next two NAWAC teams in the valley—Charlie and Delta—would be plagued by foul weather, equipment malfunctions, illness—and little in the way of definitive ape activity. Echo Team arrived on Friday, May 24, with Daryl Colyer, Travis Lawrence, Steven Matthews, Mark Dollens, and the cabin owner. That night they investigated some perplexing "meow-like" sounds;

it should be noted that in all the years the NAWAC has spent in the valley, there has never been a single sighting of a feral cat.

The next day Colyer located a fresh track 50 yards north of the east cabin. The track was 14 inches long and located behind a large tree. Later, Lawrence came upon a large flat rock with crushed nut shells and a "hammer stone" similar to what members of Bravo Team had discovered a few weeks prior.

At 9:45 p.m., the sound of a "raspy, high-pitched scream" erupted from the west. The scream was followed in short order by the sound of a tree limb snapping to the east and a wood-knock from the west. At 9:58 p.m., the team heard a "weird scream" from the vicinity of the west cabin that was immediately followed by the sound of a rock slamming into metal. Over the next two hours, the night was interrupted several times by metallic "double bangs" and "high-pitched, strange vocalizations" from the west. "They want us to chase them," remarked Lawrence.

Early the next morning, the team was greeted by the sounds of rocks striking the metal roof of the cabin. At 10:20 a.m., the cabin owner climbed up on the cabin roof to look for rocks. Just after climbing down, a rock loudly struck the roof of the east shed just a few feet away. This activity prompted Steven Matthews to make a quick patrol of the compound on his ATV. While near the west cabin, the ATV was struck by a rock. Matthews was lucky the rock had hit the machine and not him.

By the 27th, most of Echo Team had left the valley; only Ken and Ed Stewart remained. The next night, around 9:47 p.m., the Stewarts were sitting quietly just south of the main cabin and facing north when they heard a peculiar vocalization. The sound, "multiple syllables that sounded like someone talking," lasted two to three seconds. The men scanned the area with thermal units but saw nothing. Ten minutes later, they heard a similar vocalization to the east. "If I had heard that sound in a state park," Ed Stewart said, "I would have thought it was another person talking at a neighboring campsite." Of course, the pair was fully aware that all the other cabins in the compound were currently unoccupied. The men would note multiple rock-strikes and clacks until they retired to the overwatch tent at 2:30 a.m.

The next day was a rainy one. The two investigators were sitting on the east porch facing down the trail to the bottleneck when they heard a clear wood-knock ringing out from directly north of the cabin up on the mountain slope. A minute later, the sound of an approaching vehicle reached the men's ears. The cousins decided that the wood-knock was likely announcing the arrival of another vehicle in camp. In short order, Brian Brown, Alton Higgins, Jeff Davidson, and David Myers* arrived at the compound. Ken and Ed Stewart, fearing the creeks would soon become impassible due to the heavy rains, handed over control of the camp to Foxtrot Team and left the valley at 11:00 p.m. that night.

The next day at 11:40 a.m., the rain had finally stopped, so Brown and Davidson hiked over to the south cabin and planted a Reconyx camera inside one of the dead freezer units sitting outside the structure. Brown thought the odd "thumps" and "thuds" that almost every team reported were the result of apes picking up and slamming the lids on these freezers. His hope was that if an ape did open up the freezer lid, the movement would trigger the Reconyx and the group would have a photo of the culprit responsible.

At 12:30 a.m. on Sunday, June 2, the team heard a loud "whump" sound from the south. Higgins had heard the sound many times before and was convinced they were being produced by wood apes, perhaps by dropping a large rock—basketball sized or larger—onto soft soil. On this night, however, he hoped Brown was right and it was the sound of one of the freezer lids being slammed. Everyone crossed their fingers and hoped that since the sound had come from the area of the south cabin, a photo of the "whumper" might be available.

The team turned in for the night at 1:10 a.m. Higgins and Myers slept on mats on the tent floor; Davidson slept in his cot tight against the west wall of the tent; Brown was bedded down in the base cabin. At 2:30 a.m., Davidson was still awake. As he lay on his side with his back to the tent wall, he felt something "poke" him "hard" in the hip through the tent wall. Startled, Davidson moved away from the wall. After calming down, he was convinced that something—"a knee, elbow, hand, finger, something"—had poked him through the tent wall. He decided not to wake Higgins and Myers and somehow managed to lie back down and doze off.

Kathy Strain and Monica Rawlins arrived at camp at 1:00 p.m. the next afternoon. An hour later, the group heard a "prominent thump" from the north. Higgins realized this sound could not have been caused by slamming freezer lids as it had come from the mountain slope and not the south cabin area. Befuddled, but undeterred, the men decided to hike to the east and show Myers the suspected nut-crushing station Brown had discovered a few weeks prior as part of Bravo Team. The men had not been gone long when Strain and Rawlins heard a "slam" from somewhere near the south cabin. Both women believed the sound had been produced by something or someone opening and then slamming the lid of one of the dead freezers outside the cabin. The slamming sound was quickly followed by two different "rock-clacks" from the north slope.

By 2:11 p.m., the men had run into a roadblock on their way to the suspected nut-crushing station. The creek had risen to the point that it was right up against the mountain slope. The trail was covered by raging, fast-moving water. The team had no choice but to turn around. On their way back to the cabin, the men found a piece of cut firewood about 200 yards east of the compound. They felt it had been part of the base camp firewood supply and that something, or someone, had carried it to this location. After hearing of the slamming noise near the south cabin, the men went to investigate. They had barely started that way when they stumbled on yet another piece of cut firewood near Higgins's tent. This piece of firewood was pine, not the hardwood the cabin owner preferred. Puzzled, the men continued to the south cabin.

Once there, they checked the camera in the freezer. It had taken no photos and did not trigger when the freezer door was opened. Chastened, the men decided to find out if a slamming freezer door could indeed be heard all the way back at base camp. They radioed Rawlins and Strain and told them to listen. They quickly determined that anyone at base camp could clearly hear the freezer lids if they were slammed. In fact, the women reported being able to hear the freezer lids when they were dropped from a height of only three to four inches. Brown took the camera out of the freezer and placed it on a tree overlooking the dead appliances.

After investigating the south cabin, Higgins—still bothered by the discovery of the pine firewood—decided to check the wood piles of the other three cabins. As he suspected, none of the cabins had any pine mixed in with their firewood. The pine log had to have come from somewhere other than one of the cabins in the compound, but where? And how? The incident left everyone scratching their heads.

At 4:30 p.m., back in camp, Higgins discovered a "small, crushed branch" behind his tent. By stepping on a portion of the branch that had not been crushed, Higgins determined that it would have taken an animal of "quite a substantial weight" to pulverize it. Higgins thought that maybe an ape had been skulking around their tent the night before. This prompted Davidson to share his experience of being poked through the tent wall. While he could not prove it, Higgins was now convinced an ape had, indeed, visited the tent the previous night.

Just after 7:00 a.m. on the morning of the 4th, Brown, Davidson, Higgins, and Myers loaded up the truck and started out of the valley to take Myers to the airport. As they passed in front of the west cabin, Brown observed a "tall, brown upright figure" at the south fence of the property near the parked RV. He was the only one to see it; the others were not looking in that direction. "The figure was twice as tall as the fence and quickly moved to conceal itself as I accelerated to pull even with the RV," said Brown. Upon reaching the RV, Brown jumped out of the truck in pursuit of the figure. The others helped Brown look about, but in short order returned to the truck to get Myers to the airport in time for his flight.

When Brown, Higgins, and Davidson arrived back at camp at 4:20 p.m. they walked over to examine the freezers near the south cabin. The rocks, which the men had placed on top of the lids so that they could later tell if they had been opened, were gone. Excited, Brown checked the camera, but the only photos taken were of team members. Davidson would leave the valley for home shortly thereafter.

The team stayed up late broadcasting various vocalizations and producing tree-knocks in an effort to entice the resident apes into revealing their presence without luck. They finally relented and called it a night at 1:00 a.m. But at 3:30 a.m., Strain was jolted awake by a "loud bang" on the roof of the front bedroom, followed just 15 minutes later, by a rock striking the roof of the cabin above the kitchen and other noises. As Strain was making note of the various events, she detected a shift in the cabin in one direction and then slowly back. "It was as if something of significant weight or strength had pressed against the cabin wall on the east side," she wrote in her notes. Strain reported hearing two "thud" sounds outside the east wall at 5:30 a.m. that she felt were likely the footfalls of a heavy individual.

The next morning team members climbed onto the roof of the cabin to search for rocks but instead found a nine-inch-long piece of firewood. Strain felt certain that this was the object she had heard land on the roof at 3:30 a.m.

The next day was active with "lots of rustling and potential rock-throws." The rock-throwing intensified after Brown began broadcasting the Ohio Howl. At 12:32 a.m., they heard an "odd vocalization that sounded like a human scream" from the southwest. Strain whispered, "That's not good…"

The team hit the rack at 12:55 a.m. As Brown headed for his tent, he heard three heavy, clear footfalls retreating from him to the south in the area behind his truck. Brown drew his sidearm and advanced toward the sounds. Whatever it was continued to move farther off into the woods. As Brown reached the tree line, he heard some "low whispered gibberish." He quickly notified his teammates who helped him investigate the area. Multiple rocks ripped through the canopy and landed in the area of woods where the team members were looking around. At 2:10 a.m., having found nothing and running on fumes, the team again attempted to retire.

Brown had been in his tent for only a minute or two when he realized he had left his pistol on a chair back at the cabin. When he left his tent to retrieve it, he heard the sound of a large tree crashing to the ground on the slope north of the cabin. The team quickly got up and began scanning the mountainside with white lights. Brown and Strain both spotted eyeshine briefly before rocks began to rain down on the cabin compound. The activity continued until 4:00 a.m.; it ceased almost immediately once the team was inside, when they were finally able to get some rest.

The next morning Strain noticed a limb bouncing wildly up and down near the bottleneck area. She then spotted what she described as "an animal bigger than a cat but smaller than a large dog." The animal jumped from the end of the moving branch to another tree. "The animal was dark brown and had no tail," she added. While Strain could not say what the animal was, she was positive it was not a raccoon, opossum, or bird.

On June 8, the operation was turned over to early arriving members of Golf Team (the cabin owner, his girlfriend, and holdover, Kathy Strain). Rick Hayes, Jerry Hestand, Mark Porter, and Travis Lawrence were due to arrive later that day.

The week that followed would prove to be very active. The investigators were bombarded by both rocks and odd sounds over the first few days. On June 10, just a few minutes before 10:00 p.m., Kathy Strain and Jerry Hestand heard odd vocalizations from the area near the west cabin. "It was kind of a huffing sound," said Hestand. "A huff...hu...huff...hu...hu sound." Over the next several hours the team would record "an unprecedented number of rock-throws." By the time the team retired at 2:30 a.m., they had logged a total of 60 rock-throws/strikes in and around the compound, many of which were captured on the TASCAM audio recorder. Rick Hayes wrote in his journal, "Don't these apes ever rest?"

Thursday, the 13th, proved to be another day filled with near continuous rock-throwing on the part of the locals. One ominous change was that on several occasions, team members actually seem to have been directly targeted. At different times, Hayes, Strain, Porter, and Hestand each had rocks land practically at their feet or bang off structures they were standing next to with substantial force. Had more aggressive rock-throwers joined the party or had the resident apes simply grown frustrated by the extended stay of NAWAC team members?

Rock-throwing continued in earnest that night. Around midnight, Strain was fortunate to escape injury when two rocks whizzed by her head and struck the cabin wall behind her. These two rocks were quickly followed by two more that landed on the roof of the cabin and east shed respectively. How could the apes remain hidden yet still be so accurate with their throws? The answer would remain as elusive as the animals themselves.

Rock-throws, wood-knocks, and the sounds of multiple animals thrashing about in the brush on the slope of the north mountain continued into the wee hours of Friday, June 14. At 1:52 a.m., exhausted and frustrated, Hayes and Lawrence began to "return fire" by picking up and hurling rocks of their own up onto the mountain. The locals answered by pounding the base camp cabin multiple times. At 2:15 a.m., Hestand reported hearing a huge commotion from the vicinity of the south cabin. "The apes are going ape over there!" he yelled to Lawrence and the others. They heard the sound of metal being banged upon and what was believed to be the lids of the dead freezers being opened and slammed repeatedly. At 3:00 a.m., despite the barrage continuing all around them, the team retired to the cabin. While his teammates tried to sleep, Lawrence established an overwatch position in the front bedroom, but he never caught a glimpse of the hurlers.

The next day Porter and Hestand headed home. Around 8:00 that night, after hiking the ATV trail talking and using white lights, Hayes and Lawrence returned to the base camp, which triggered a new rock bombardment. The rocks seemed to be coming from all directions. Needing to release some steam, Hayes and Lawrence began firing rocks into the middle woods and onto the mountain slope. Suddenly the team heard an unnerving vocalization from the middle woods area. "It was a deep, growling, almost maniacal sounding pant-hoot," said Lawrence. "It was a ma...ha...ha...ha...type of sound." The sounds of rocks striking the east cabin and the overwatch tent continued over the next several hours.

On the morning of the 18th, after enduring a strong thunderstorm during the night, Kathy Strain noted in her field journal that "No nuts or other sounds were heard on the roof overnight." The lack of impact sounds—even during a heavy storm—bolstered Strain's belief that what she and others had been experiencing were actually thrown projectiles and not falling nuts or debris from the nearby trees.

At 3:09 p.m. on Thursday, June 20, the team was startled by what they described as a "huge, metal sound" originating somewhere near the south cabin. Investigation yielded nothing. Burrows and Lawrence decided to conduct a reconnaissance of the compound. Burrows hiked to the west cabin where he hunkered down in a spot about 35 yards south of the building and just north of the creek. Lawrence hiked up the mountain and proceeded to walk in a large east-to-west loop around the compound until he met the main road. At that point, he turned back to the northeast and headed toward Burrows.

As Lawrence was attempting to locate Burrows, he saw two large gray legs "scissor" as the animal walked through dense foliage 30-40 yards southeast of the west cabin. Lawrence was momentarily confused and thought he was seeing Burrows. He then saw another set of gray legs (a second animal) moving bipedally through the thick brush. Both animals moved to the south and away from the two investigators. Lawrence then met up with Burrows who had not seen the apes. The pair attempted to follow the two animals but could not pick up their trail.

Between 7:15 p.m. on the 20th and 2:00 a.m. on the 21st, the rock bombardment continued. Late that morning, Strain saw a rock hit the canopy roof of the overwatch tent from the south-southeast. The harassment was no longer strictly a nighttime event; the valley residents were now boldly and defiantly launching their assault in broad daylight. The team remained perplexed as to how the culprits could move in such close proximity to camp without being seen. The rock barrage continued all night and into the wee hours of the 22nd.

Mark McClurkan, the first member of India Team to arrive, pulled into camp at 2:03 a.m. Rather than disturb his sleeping teammates in the cabin, he decided to rest in his truck. About 20 minutes later he heard movement in the brush approximately 40 feet to the south (his windows were down), followed by a sort of high-pitched chatter described as like a "eeeee bwah gee bah doh." After the chatter, something heavy landed with a thud in the brush and McClurkan—thanks to a nearly full moon—was able to catch a glimpse of something upright moving behind the outhouse. McClurkan then decided to enter the cabin and report the encounter.

The next morning, Lawrence, Strain, and Burrows, after experiencing one of the most active weeks in the history of Area X, left the valley. McClurkan remained, alone, to await India Team members Robert Taylor and Ken Helmer.

McClurkan was scheduled to spend several days and nights alone in the valley before his teammates joined him on Thursday. His first full day alone was filled with the sounds of metallic bangs, rock-strikes, what sounded like the slamming of doors (freezer lids?), and wood-knocks. Having been advised by Field Operations Coordinator Daryl Colyer not to stray far from the relative safety of the base cabin until others arrived, McClurkan chose to simply observe and record the auditory incidents rather than pursue the noise-makers.

At 8:30 p.m., McClurkan heard what he described as an "odd sound" from the mountain slope just behind the cabin. "It was almost like the moan of a woman softly crying," he said. "It was a bit disconcerting." So McClurkan decided to light a small fire in front of the cabin. While sitting next to the fire, he watched an aluminum beer can fly into camp from the west, followed by the sound of "very heavy footsteps" behind the cabin. Moments later, he heard "huffs" from the same general area. Alone and without backup, McClurkan stoked the fire and stayed put. At 9:23 p.m., he heard a "very deep howl" from about 500-600 yards away. The howl was followed by what sounded like bipedal walking up on the slope only 50-60 yards north. McClurkan, an experienced outdoorsman and tracker, admitted to feeling more uneasy than he had ever felt before in the woods. "I believe these animals absolutely know I'm here alone," he wrote in his notes.

At 8:05 p.m. on the night of the 24th, the raw-nerved McClurkan heard what he described as "squeaking sounds" from the mountain slope and odd "metallic" sounds from the direction of the south cabin. The oddest incident that night involved the burn barrel east of camp where the lone investigator had burned some trash earlier. "I heard something bumping the burn barrel fairly hard," he said. "Then the fire flared from the coals it had produced earlier." McClurkan did not think any combustible material remained in the barrel. A few minutes later, the event repeated with an even brighter fire flaring to life. Had a wood ape approached the burn barrel and tossed in some sort of tinder? If so, why?

Robert Taylor and Ken Helmer arrived at the cabin at 6:00 p.m. on the 27th. McClurkan, understandably, was glad to see them. About four hours later, the men decided to act on an idea originally suggested by investigator Alex Diaz. The plan was to hang a large white screen up in front of the base cabin and create a make-shift drive-in theater. The movie selected—for obvious reasons—was the documentary *Chimpanzee*. The results were dramatic and almost instantaneous. In short order, they heard a wood-knock, sticks breaking, movement, a rock-strike, and a "grunt or growl." And a strong pungent odor wafted into camp.

"I think it is about to go down," Helmer said to his comrades. "I bet this chimp movie drives them bananas, pun intended." By 10:25 a.m., the movement sounds had intensified and were coming from multiple directions simultaneously. "I feel like we are surrounded and about to get ambushed," noted Helmer. There seemed to be at least three, possibly four, animals in the immediate area. Taylor picked up one of the thermal units and scanned to the northwest where he reported seeing a "face" in the brush. Helmer and Taylor then observed a large, upright animal in the brush walking away from them. Within four seconds, the creature was gone.

Midnight arrived, but the new day brought no slow-down in activity. After multiple rock-throws, Helmer said, "There are freaking apes everywhere! They are on us like stink!" The men were startled a few minutes later when a "very large rock clobbered the cabin." Rocks continued to fly into camp and at 2:30 a.m., the team heard a "long, low howl" coming from the north that lasted six to seven seconds. McClurkan said, "I am absolutely certain that vocalization was not produced by a coyote." Within minutes, a pungent and foul odor drifted into camp. By 3:30 a.m., the men were too fatigued to stay up any longer and retreated to the cabin. Their rest would be fitful, however, as the cabin was periodically pummeled by rocks the rest of the night.

By noon on the afternoon of the 29th, McClurkan and Taylor would leave the valley, replaced by incoming members of Juliet Team: Daryl Colyer, Jordan Horstman, and Brad McAndrews. That night, Horstman and McAndrews documented numerous rock-strikes, a deep "hoo-hoo" sound from the mountain slope, and what the pair described as "like a human voice whisper-talking."

At 9:00 a.m., Horstman and McAndrews left the overwatch tent and joined Colyer in front of the cabin. At one point, Horstman caught sight of a small tree shaking violently just west of camp. "Something's moving that tree!" he cried. The tree was only 20 feet away. Colyer and McAndrews quickly advanced on the tree but the "shaker," whatever it had been, seemed to be gone. Horstman, who had stayed back watching, saw a "small black animal hanging by one long arm from a tree limb" before it dropped, embraced the trunk of the tree, and quickly climbed down. "It was about the size of a 50-pound bag of feed and totally black," he said. "If I had been at a zoo, I would have known I was looking at a small ape or monkey." The team spent the rest of the morning looking for, and finding, multiple rocks on the cabin roof.

On July 2, after a night filled with still more rock-strikes on and around the cabin, an exhausted Colyer went to bed shortly after midnight while McAndrews and Horstman began overwatch. At 2:32 a.m., the men heard a "sharp, loud wood-knock" from the north mountain slope. Within five seconds, a duller sounding knock was heard from the east. Only ten seconds after that, three more "loud" wood-knocks rang out from the north mountain "knocker." Less than 30 seconds passed before there was another knock from the west. The men were amazed, feeling they had been witnesses to some sort of cryptic communication.

"Amazing," McAndrews wrote in his field notes. "Six wood-knocks in 35 seconds from three locations. How is this possible? The apes must harbor the ability to plan for future events by strategically carrying tools (rock or stick) for communication. It's as if they have triangulated the area immediately around the base camp cabin.

"The last three nights have provided a treasure trove of data. Our presence in the overwatch tent may present us with the best opportunity to observe wood ape behavior in a concealed manner. Previous suggestions that the overwatch tent might alter ape behavior are unfounded, in my opinion. The overwatch tent should be viewed and utilized as a most valuable and productive asset."

At 2:37 a.m., they heard yet another wood-knock. "The apes must be communicating something," Horstman wrote in his journal. "But what?" An hour later, the men heard two wood-knocks from the north mountain. Seconds afterward, they heard the sound of a "rock-on-rock clack or a wood-on-rock knock." Wrote McAndrews: "It is clear, in my opinion that these wood-knocks are being used for purposes of discreet communication."

Wood-knocks and metallic banging sounds continued over the next two hours. Horstman, struggling to stay awake, finally gave in and lay down on the cot. But McAndrews was determined to make it until at least daylight. "We continue to observe what appears to be a series of communicable aural signals which originate from different locations," he wrote in his journal. "These 'signals' are a result of rapping an object (wood, rock, or metal) with another object used as a tool or anvil. Basically, various objects appear to be used as tools for the purpose of communication. Are they simply communicating locations to one another? If so, why? For what purpose?"

Dawn did not slow the bombardment of rocks landing on and near the cabin and its satellite structures; one later whizzed by Horstman's head, barely missing him. The barrage continued until noon.

The next day, the rock barrage began anew with multiple stones landing on or near the cabin between 7:30 and 8:32 p.m. Four rocks, thrown in quick succession, landed close to Colyer. An aggravated Colyer wrote about the incident in his notes: "It's so frustrating to pursue a flesh-and-blood animal that is so often impossibly elusive; they will not be pursued, at least not in the conventional manner. So frustrating that the moment I turn my back to them they launch a barrage of rocks my way, yet when I turn to confront them, they disappear like ghosts. No wonder the American Indians ascribed to them supernatural attributes. Though we are not engaged in any sort of military campaign, often it seems they are well versed in guerilla tactics, striking and disappearing before we can bring our technology to bear. They are at once highly infuriating and magnificent. It is no wonder they have not been officially discovered yet. We will be most fortuitous for our work to culminate with that discovery."

On Sunday, July 7, NAWAC personnel switched out again with Daryl Colyer leaving the valley and Mark McClurkan and Robert Taylor joining Alton Higgins and Travis Lawrence who had arrived four days before. Lawrence, Higgins, McClurkan, and Taylor comprised Kilo Team. The next several days featured the now common rock-throwing, several "weird, drawn-out, hoarse howls," and multiple wood-knocks. These events would be only a prelude to what might be the closest the NAWAC has come to collecting a specimen.

16

The Kilo Incident

Just after midnight on the 11th, activity began to pick up again with multiple wood-knocks and "loud noises and vegetation moving" behind the base cabin. At 1:15 a.m., the Kilo Team commenced their nighttime operations. Higgins retired to his tent and McClurkan to the cabin. Lawrence and Taylor entered the overwatch tent; they noted much activity over the next two hours. "Not 15 minutes went by without us hearing something," said Lawrence.

At 3:50 a.m., Higgins exited his tent and stepped outside to urinate. He stood on the west side of his tent facing the wooded area to the southwest with his back to the overwatch tent. That's when Lawrence, scanning with the ATN thermal scope on the .30-06 overwatch rifle, saw a huge white-hot heat signature appear beyond the tent and to the right of Higgins just inside the wood line of the southwest woods.

Drawing of what Travis Lawrence saw through his thermal scope the night of the Kilo Incident.

Lawrence watched as the creature stood up, revealing its body from the armpits to the top of its conical-shaped head. "It looked *nothing* like a human," Lawrence said. "It was huge, had a pointed head, enormous shoulders, and trapezius muscles that blended into the head which was set lower than that of a human." The creature had apparently been quietly squatting in close proximity to Higgins's tent, only 10-15 yards away, for hours.

Deeply concerned for his friend's safety, Lawrence placed the reticle of the scope on the spot where he believed the animal's nose would be located and fired. Higgins immediately dropped to the ground and crawled back into his tent. After a few seconds, Lawrence instructed Higgins to leave the tent and go inside the cabin. Higgins complied, entered the cabin, and woke up McClurkan who had slept through the entire episode.

Shortly after 4:00 a.m., Lawrence and Taylor exited the overwatch tent and met up with McClurkan and Higgins near the cabin. The four men began searching the southwest woods behind Higgins's tent for a body. They found nothing. No body, no blood. Lawrence simply could not believe it. "There is no way I could have missed from that distance," he said. "No way." At 5:00 a.m., the men decided to postpone their search until first light. Just 20 minutes later, a rock slammed the roof of the cabin. Sleep would prove tough to come by.

The next morning, the team began the investigation of what would come to be known as the "Kilo Incident." The men carefully paced off the distance from the overwatch tent to the spot where the creature had been standing and measured 30 feet. Lawrence again stated his disbelief that the ape had not dropped on the spot. "I just don't know how I could have missed it," he lamented. The words were barely out of Lawrence's mouth when Higgins called the group over to the area behind his tent.

He had found a broken tree limb, about the diameter of a pencil, that appeared to be in the flight path of the bullet. McClurkan examined the limb, returned it to its original position, and observed a "cup-shaped notch" on the top of the limb where the bullet had contacted it. McClurkan then found a second limb, a bit higher than the first, that had been hit and displayed a similar notch. What had happened suddenly became clear to the seasoned trapper and hunter. McClurkan explained that the bullet had been deflected upwards and away from the face of the wood ape by the first branch. If the trajectory of the bullet had been one-quarter inch lower, things would have worked out much differently. The other three men could only shake their heads in disbelief. "How lucky can these things be?" Lawrence asked.

Higgins wasted no time in recording details of the Kilo Incident while they were still fresh in his mind. He wrote: "The overwatch tent is about 52 feet from the base cabin. The overwatch tent is about ten feet square. It is set up to house a 'scanner' (a team member employing a hand-held thermal viewer) and a 'shooter' (armed with a 30-06 equipped with a thermal scope) The 'snorer' sleeps in a small tent located 50 feet west/southwest from the overwatch tent in the general direction of the west cabin.

"The bullet fired at the figure early Thursday morning hit a twig approximately 6'4" high, fifteen feet from the front of the tent. The bullet exit hole in the overwatch tent's plastic bag wall was approximately 3'10" above ground level. The bullet rose 30 inches from the overwatch tent to the twig that deflected it, or about 0.448 inches per foot. Adding eight inches to account for the distance from the nose area—where Lawrence aimed—to the top of the head (which I think is a very conservative guesstimate), and adding another four inches to account for the low spot where the subject stood, and adjusting for the continued rise of the bullet, produces and estimate of the subject's height of 7'10" to 8'2".

"Lawrence had McClurkan stand where the subject stood while he observed from the overwatch tent. McClurkan is about 6'1" tall. Lawrence estimated the subject's height at 8'7" based on this comparison. The observed and measured flight path of the bullet, including rise and distance to target, serve to corroborate Lawrence's account and his estimate regarding the size of the thermal image subject. Because the correction factors were believed by me to be conservative, it could very well be that the subject had a height of 8.5 feet or more."

Around noon, the team conducted an experiment to see if the plastic wall of the overwatch tent had altered the trajectory of the round fired the previous night. The men set up a heated can (so the thermal unit would pick it up easily during daylight hours) and set it up 40 yards from the overwatch tent. Lawrence hit the can on the first shot. It was obvious that the plastic had neither deformed the bullet fired the night before or altered its trajectory. "That's the luckiest ape that's ever lived," mumbled Lawrence.

After much discussion that day, the investigators decided to recreate the scenario that had led to the shot opportunity the night before. At 1:25 a.m., Higgins retired to his tent. McClurkan went inside the cabin and Lawrence and Taylor began overwatch. Just like the previous night, no more than 15 minutes would pass without there being some kind of auditory event. As per the pre-determined plan, Higgins exited his tent at 3:50 a.m.—just as on the previous night—and stood as if urinating. Through the handheld thermal, Lawrence observed what he was certain was the heat signature of an ape rising up from behind the brush to observe Higgins. Amazingly, the figure was in the same spot where the ape had stood the night before. Lawrence alerted Taylor, the rifleman that night, as to what he was observing. Taylor attempted to locate the subject but could not find it. By the time he handed the rifle to Lawrence, the animal had disappeared. The rest of the night was uneventful, and the pair ceased overwatch at 6:30 that morning.

The following weeks would be fairly tame by Area X standards. The next event of significance occurred on June 23. Ken Helmer, Paul Bowman, and his friend, Lee Bailey* were the investigators in camp that day. At 2:50 p.m., Bowman, in the hopes of eliciting a response from the local apes, began beating on a sheet of corrugated metal with a baseball bat. As he did so, Bailey hiked about halfway up the mountain slope. At 3:30 p.m., Bailey heard Bowman start a second round of metallic bangs down at base camp. His attention quickly switched to the sound of an approaching animal. Bailey decided to sit down on a log and wait for the animal to emerge; he was certain it was a white-tailed deer. No deer emerged from the brush; instead, Bailey observed "two charcoal-colored human-like legs from the knees down" through a gap in the vegetation. He watched the legs move as the animal retreated from his location. The creature picked up speed, and Bailey could hear it crashing through the brush as well as its "heavy-thudded footfalls." Stunned, Bailey returned to camp and shared his encounter with Bowman and Helmer.

After an uneventful week where the cabin owner and his son were the only people in camp, Alton Higgins and Travis Lawrence returned on August 6. The next day, the men decided to implement an idea of mine. I suggested creating a "hide" or "blind" from the same black plastic that formed the walls of the overwatch tent. The wall would be placed near the camp in a spot where apes had approached before and which would be visible from the overwatch tent. My theory was that an ape attempting to observe the team might utilize the black plastic wall as cover, not realizing the men conducting overwatch could see them with the thermal units. I posited that if an ape hunkered down behind what was dubbed the "Mayes Wall" and felt it could not be seen, it might stay in place long enough for a shot to be taken. With all that in mind, Lawrence and Higgins constructed a wall 17 feet long by about 5 feet high near the cabin and within the direct view of anyone in the overwatch tent.

At 6:40 p.m., Higgins erected his tent in his usual location. Then, after supper, the pair double-checked to make sure Lawrence would be able to see through both the overwatch tent wall and the Mayes wall plastic. Lawrence reported that it was "easy" to see Higgins behind the wall through the thermals.

Unfortunately, the two nights of overwatch that followed did not yield any visuals. The next couple of weeks would prove to be unusually quiet.

Alton Higgins and Daryl Colyer, the first members of the NAWAC's 17th and final team of the summer, Sierra Team, arrived at the compound property gate at 11:25 p.m. on Friday, August 30. The men found the property gate closed and locked. The pair would have to hike to the cabin in order to retrieve the key needed to unlock the gate. Upon reaching the cabin, the pair found that the overwatch structure had been flattened. Upon investigation, Higgins found that the structure had not just been knocked down; rather, the metal legs and frame had been snapped into pieces. While Colyer and Higgins were discussing the damaged overwatch structure, an explosive metallic bang erupted from behind the cabin. "It sounded like a lightning strike," Colyer said.

The men hurried to the area behind the cabin and scanned the slope for movement and/or eyeshine. They saw neither. Chastened, the men returned to their original task: retrieving the gate key from the cabin. As the men walked around to the front porch of the cabin, they saw that it had been completely ransacked. All manner of gear and equipment was strewn about. The men figured a raccoon or wayward black bear was likely the culprit and retrieved the gate key. The rest of the team—Travis Lawrence and Ken Stewart—arrived at 1:03 a.m. Metallic bangs, a "hmph" vocalization, footfalls, and rocks landing on the cabin were heard at intervals throughout the night.

Lawrence and Stewart conducted overwatch during the early morning hours of September 1. At 1:25 a.m., Lawrence clearly heard the sounds of "feet slapping the ground" as something large and fast sprinted down the road between the base cabin and the west cabin. This was followed by a wood-knock from the west and a sharp "metal bang" from the east. The men ran out of steam at 4:30 a.m., aborted overwatch, and retreated to the cabin to sleep.

At 6:19 a.m., Higgins—who, as always, was sleeping in his small tent—"sat bolt upright" upon hearing what sounded like soprano singing voices from the woods. Everyone else was still dead asleep. "I am dumbfounded," he wrote in his journal. "Angel choir?" He then wondered if this might be some sort of response to the haunting Gregorian chants Colyer had played in camp the night before. "The 'singing' was multi-tonal and sounded as though more than one voice was producing the sound. I am stupefied," wrote Higgins.

When, later that morning, Higgins told his teammates about the "singing," Stewart reminded the group that he had put out an audio recorder before going to bed. Stewart retrieved the recorder and found that it had indeed recorded the very faint sounds so reminiscent of a soprano choir at the exact time Higgins had claimed to have heard the "singing." The men were absolutely flabbergasted. What kind of place was this?

Several uneventful days followed. The men woke up early on the morning of the 3rd and began cleaning up the mess created by whatever had ransacked the porch. Initially, the men had assumed a raccoon or a black bear had been the culprit. But as they cleaned up and inspected the various items strewn about, they began to change their minds. There were no bites or claw marks of any kind on any of the equipment or canned goods. In fact, none of the food items had been eaten. "It was as though someone, or something, had ransacked the porch items and threw them all about simply for the fun of it," remarked Colyer. Among the untouched food items was a crushed can of Pringles and a loaf of bread. None of it had been torn open. Certainly, the men figured, a bear or raccoon would have eaten the food items.

At 9:45 a.m. on September 3, Colyer and Higgins left the valley. Operation Relentless had ended. As usual, the men left with more questions than answers.

17
Operation Tenacity

Prior to the beginning of the NAWAC's fourth prolonged field study in Area X, the group undertook a large-scale renovation of the northern base cabin. Most importantly, the group added an elevated wooden observation deck to the northeast corner of the structure. The ten-foot square platform had black plastic "walls," and was topped with a canopy. Dubbed the "Overwatch Tower," this new elevated blind provided a nearly 360° view of the camp and would be the location from which members would conduct surveillance during nighttime hours.

Operation Tenacity kicked off on May 17, 2014, when Alpha Team members Mark Porter and Bill Coffman rolled into camp. Field Operations Coordinator Daryl Colyer arrived at 2:38 a.m. and made no effort to enter the cabin compound quietly. Instead, he honked his horn relentlessly and yelled out the window of his vehicle. "My hope was to disrupt the quiet night and raise the anxiety level of any apes in the area," said Colyer. Upon arrival, he yelled, "I'm here," made a quick reconnaissance of the compound, and scanned the slope of the north mountain with white light in an effort to "shake things up." All was quiet, so Colyer turned in for the night. When he turned out his light, the wood shed on the west side of the cabin was struck loudly by what he figured was a rock. "Mission accomplished," Colyer thought before going to sleep.

Over the next several days, the team documented multiple wood-knocks, anomalous thumps, rock-strikes on the cabin, and located a likely nut-crushing station south of the cabin. They found four "hammer stones," all with the remnants of crushed black walnuts adhering to them. Numerous walnut shells littered the ground around the largest stone. The men looked around carefully but found no black walnut tree anywhere in the vicinity.

Bill Coffman left the valley on the 21st, but the team was reinforced by the staggered arrivals of Ken Helmer, Ed Harrison, and Rick Hayes. At 4:46 p.m., a powerful wood-knock rang out from somewhere east of camp. Immediately following the loud knock, Hayes heard what he described as "strange chatter that sounded like faint, indiscernible human speech." Investigation yielded nothing. Overwatch that night provided no visuals; however, there

155

was no shortage of anomalous sounds. Rock-clacking, a "loud double impact" on the roof of the west shed, and "eight successive metal bangs from the west cabin area" were documented.

At 11:15 on the night of the 22nd, Colyer again attempted to stir up the apes. He walked west to the point where the trail into the cabin area meets up with the main compound path and began yelling, screaming, beating his chest, and imitating an angry chimp's pant-hoot call. His vocalizations echoed through the valley and, no doubt, carried quite a distance. A mere 30 seconds later, a rock flew down from the mountain slope—clipping leaves and other vegetation on its way—and landed just short of the cabin. Colyer, Helmer, and Harrison all heard the rock land and bounce away. Just moments after the rock-strike, Colyer heard a series of peculiar vocalizations the likes of which he had never before encountered. "They were low in tone, gravelly or raspy; they were growling, grunting, and chattering in nature," he said. "There were four vocalizations in all." Colyer attempted to mimic the odd vocalizations but heard no reply and returned to the cabin.

Just after midnight on Friday, May 23, Hayes and Harrison entered the overwatch tower and did not have to wait long for the activity to start. First, a couple of rock-strikes to either the west cabin or the RV, then the sound of "something metallic being moved around" coming from the west. At 3:50 a.m., Harrison, facing east toward the bottleneck area, spotted a white heat signature through his scope. He watched the figure as it moved down the slope and the edge of the tree line. Much to his chagrin, the creature did not step into the open; instead, it appeared to squat down and hide behind some brush. Harrison could now make out only the top of the heat signature. Although convinced he was watching an ape observing base camp from a distance of only 50 yards, he held his fire as he could not be absolutely sure. The animal stayed in place for only a few minutes before it faded back into heavier brush where it was totally concealed. Harrison did not see it again. Nothing else of interest was seen or heard that night.

Bravo Team, made up of Brad McAndrews, Travis Lawrence, Phil Burrows, Brian Brown, Jeff Eltringham, and Tom Stevens* arrived on the 23rd and joined Alpha Team holdovers Colyer and Helmer. Also, in attendance as a special guest of the NAWAC was wildlife biologist David Myers.

Over the next two days the investigators heard "a low moan vocalization," a sound "like a can being kicked," "two shrill and high-pitched" screams, and rock-impacts. In the early morning hours of the 26th, Brown broadcasted the Ohio Howl multiple times. There was no immediate response, but later the overwatch team heard a "whoop-like" vocalization at 4:19 a.m., followed by "an odd sound, possibly a vocalization" that was reminiscent of a "wind-blown pipe-like sound." The rest of the night was quiet.

Upon returning from a hike on the afternoon of the 24th, David Myers heard a "faint two-note whistle" to the north. It was a faint, high-pitch to low-pitch sound similar to the tone someone might use when calling out "yoo-hoo." Myers responded with a two-tone whistle of his own. He was rewarded when the unseen whistler answered. Myers again attempted to elicit a response with a whistle of his own. The unseen source took longer to respond the second time but eventually did answer, albeit from much farther away than the first two times. Myers initially commented that he did not feel the sounds were avian in origin but finally decided the whistles must have been produced by a bird.

Pouring rain set in and seemed to subdue wildlife activity until the afternoon of Wednesday the 28th when the team heard a "clear wood-knock" followed by a "very, very loud impact sound." At 6:20 p.m. McAndrews began firing test rounds from the Marlin .45-70 as he believed the overwatch weapons and optics were out of alignment and needed "zeroing." Minutes later, the team heard a sharp rock-strike on the west RV. As McAndrews continued his efforts to zero in the .45-70, they heard another rock-impact on the slope to the northwest. At 6:45 p.m., the investigators heard loud cracking noises from a tree to the northwest of the cabin. Burrows and Brown, who was stationed at the base of the slope behind the overwatch tower, saw the tree swaying violently. Myers was on the trail farther to the west and also saw the swaying tree.

Thinking the tree was about to fall, Brown ran into the cabin and retrieved his video camera. As he exited the cabin and came around the northwest corner of the building with the camera running, he heard Burrows yell, "There's something in the tree!" Brown panned up and repeated Burrows's statement, "There *is* something in the tree!" The tree began to fall and Brown yelled, "It just jumped out of the tree!" He heard something land with a dull thud, presumably an animal of some kind, and then loud movement away from the cabin to the west. The tree then fell over with a tremendous crash. Eltringham called out, "How big was it?" Brown replied, "It was fuckin' big!"

The incident troubled Myers greatly. He really wanted to attribute the incident to a known animal but struggled to do so. Myers was "impressed" with what he witnessed during his stay with us, but he was always very hesitant to use the term "Bigfoot" or "ape" as he had never seen one. While open to the idea, all he usually offered was what he knew the noises heard were *not*.

Afterwards the men wondered if an ape had taken the loud rifle shots from McAndrews's .45-70 as some sort of threat. Actually, they questioned whether or not an ape had been involved at all. After all, only the dark blur of some kind of animal was seen—and videoed. All that could be said with certainty was that a large animal of some kind came down out of a tree not capable of supporting its bulk. Still, the rock-throwing incidents in the minutes prior to the tree falling led them to believe that an ape had been the culprit.

At 6:21 a.m. during the night overwatch, McAndrews, who was alone, reported hearing "two very odd vocalizations to the south, like a cross between a snort and a whine." It was repeated just four minutes later. "The vocalization was reminiscent of a velociraptor from the *Jurassic Park* movie!" McAndrews wrote in his field journal. "It sounds like a damned dinosaur. No kidding. I am not hallucinating and am of clear mind. I am not tired." There would be no more activity that night.

The first members of Charlie Team—Mark McClurkan, Gene Bass, and Justin Horn—arrived in the valley in the early afternoon of Saturday, May 31. They didn't have to wait long for the excitement to begin. At 4:00 p.m., McClurkan spotted an "upright, reddish-brown figure" moving west to east across the trail that led to the south cabin. Neither Horn nor Bass saw the figure, but they did hear the sound of a large animal moving through the brush.

At 12:45 a.m. on Sunday, June 1, McClurkan was reading an ebook on his phone and had the fire ring to himself save for the camp fox that kept close in order to snatch bits of flour tortilla from the investigator from time to time. Without warning, there was "large movement from every direction and in very close quarters." The fox immediately fled to the east. Unable to see the animal(s) in question, McClurkan retreated to the safety of the cabin. At 3:00 a.m., McClurkan heard what sounded like "clear, bipedal footsteps" that stopped, seemingly right at his window. When he heard some "very heavy breathing." McClurkan picked up his sidearm and sprang to his feet to face down the window-peeker. But McClurkan only heard the sound of steps quickly moving to the north.

At this point, you might be thinking, *That McClurkan guy sure has a lot of sightings.* Later, you may find yourself having similar thoughts about investigators Travis Lawrence, Daryl Colyer, and others. The fact that these men had more visuals than other NAWAC members is directly attributable to the amount of time they spent on site in Area X. McClurkan, for example, was likely second only to Travis Lawrence with regards to the number of days spent in the valley during those first four years. As a result, he had quite a few startling experiences.

There is a direct correlation between the amount of time spent in the valley and sightings or other anomalous experiences. The more time you spent in Area X, the more likely you were to see something. It's also worth noting that sightings often occured in "waves" or "flaps." Often, there will be two to three weeks of intense activity followed by a month or more of little to no activity. If an investigator was on site during one of these active periods, it is only natural that he/she would have more visuals. Although this narrative might give you the impression that the activity was non-stop in Area X, that was simply not the case. In fact, most of the time nothing, or very little, happened.

Shortly after Shannon Mason and Shannon Graham arrived at the compound just before 8:00 a.m. they heard a "weird, siren-like vocalization" erupt from the west woods. It was unlike anything any of them had heard before. That evening the team documented two rock-strikes on the west RV, a "short, very high-pitched yell," a "whoop," also from the west, and animal movement in the woods to the southeast of camp.

Only half an hour into Monday, June 2, several rocks rained down on the ground behind the cabin and movement was heard in the middle woods. As the team investigated, Graham reported seeing a pair of glowing eyes seven to eight feet above the ground. "They blinked," she later said.

Encouraged, McClurkan devised a plan he hoped would draw an ape out into the open. Initially, three investigators would walk down the trail together. Then two investigators would hide in the woods on the north side of the trail, while the third investigator, with headlamp on, would continue walking in plain view down the trail. The hope was that if an ape was watching, it might have observed the one headlamp and assumed only one person was walking the trail. If so, it might approach the two hidden investigators as it sought to follow investigator with the headlamp back to camp.

At 12:55 a.m., this exact scenario seemed to play out as both hidden investigators began to hear something big creeping through the brush very close to their locations. The sound of "heavy, bipedal movement," said Graham, "was close enough that I felt I could have reached out and touched it." Despite this close proximity, she was not able to see the creature in the inky blackness. Understandably unnerved, Graham called out to McClurkan, who clapped his hands loudly in response. The movement ceased, and the pair hiked back to the cabin.

When asked why McClurkan had not lit up the animal and attempted to collect it, he replied, "It simply got too close too quickly. A missed shot or just scaring it could have turned out badly. It was best just to turn it at that distance. At 30 feet it would have prompted a shot; at less than ten feet...I just wasn't comfortable going there."

The next day passed with only a few unusual sounds reaching the ears of the team, but the night would prove to be very different. At 1:15 a.m. on Tuesday, June 3, McClurkan and Mason were walking down the trail toward the west cabin when they heard the sound of a large animal moving through the woods to their right. Within seconds, they were overwhelmed by a "very powerful zoo smell." After a pause, the two investigators continued their night hike to the west, when suddenly, McClurkan observed, just feet in front of him, a "tall, dark, upright shadow" rise up and dart across the trail from south to north. "Back up! Back up!" he hissed to Mason. Moments later, the pair cautiously made their way back to the main cabin.

Adrenaline pumping, McClurkan and Mason grabbed lights and night vision equipment, which they regretted not having had with them from the start, and quickly returned to the site of the incident minutes earlier. Seeing nothing with the night vision, they scanned the surrounding area with white lights. As they swept the fenced area near the west RV, both saw two large eyes reflecting their light. "They blinked, looked to the right, and then dropped" out of sight. Nothing else was seen that night.

At 11:30 a.m. on Tuesday, June 3, Graham and Mason were hiking where the trail widens at the bottleneck area, when Graham watched as a large, stooped, brownish-red figure dart across the trail from south to north toward the mountain slope at a distance of no more than 50 yards. Graham was amazed at the smoothness of the animal's movement. "It seemed to glide," she said. "It was gone before I could react." Mason, who had been about five yards behind Graham, did not see the figure cross the trail. But when told of the event, she drew her sidearm and sprinted toward the spot where the animal had fled. "There's no way in hell we can catch it," Graham said.

It was daylight. Why did no one videotape the event, you might ask? Because this was a collection team. They were carrying guns, not cameras. Again, the main goal was a specimen, not photos or videos.

The last significant event to take place on Charlie Team's watch was when "something powerful" struck the cabin's back wall, causing the entire structure to shudder. The incident occurred at 10:45 p.m. on Friday, June 6. Despite investigation and Bass scanning the slope with a thermal, they saw nothing. Charlie Team would pack up and leave the compound the next morning.

Just before 5:00 p.m. on Saturday, June 7, I, along with Dusty Haithcoat and Tony Schmidt, the first members of Delta Team, arrived in the valley. We were all very excited as we had been briefed on the incredible level of activity over the first three weeks of the operation. But my hopes of a week of ape activity were dampened—literally and figuratively—by a positively massive rain event that set in on the valley that first night and did not cease until the afternoon of the 10th when Travis Lawrence and Alton Higgins joined us.

Around 11:40 a.m. on the 11th, Lawrence, Higgins, and I were descending from top of the north mountain when Lawrence caught a glimpse of a "long, low, tawny-colored animal slinking through the brush." He was sure he had spotted a mountain lion. That would be all for the excitement, as it would rain all night until the next afternoon when Higgins reported seeing a "light-colored something move like a man" while coming down the mountain. Thinking that he may have seen his teammates who were around on the slope, Higgins then realized that teammates were all wearing dark or camouflaged clothing. No one was wearing anything light-colored or gray. The consensus was that Higgins might have caught a glimpse of "Old Gray" himself on that slope.

The members of Echo Team—Jerry Hestand, Mark Porter, Marvin Leeper, Ken Stewart, and Ed Stewart—joined holdovers Higgins and Lawrence in the valley by the afternoon of June 14. The next day the team documented wood-knocks, rock-strikes, and a "soft three to four syllable garbled growl." The team dutifully investigated each odd sound but found nothing. At 5:59 p.m., the men were returning from setting up some Plotwatcher cameras along a nearby game trail when they heard a "LOUD falling limb (tree?) crash" from up on the mountain slope. Hestand and Lawrence rushed to investigate while Higgins scanned the mountainside with binoculars. Higgins caught a brief glimpse of a "smallish black figure quickly moving from east to west and angled slightly downhill." The animal moved quadrupedally, "like a chimp," but exhibited longer arms than legs. Higgins was adamant that what he had observed was not a bear.

A bit later, around 6:30 p.m., Higgins was approximately 40 yards west of the west cabin and the parked RV when he heard "three or four crashing sounds like running footsteps" ahead of him. Higgins then saw a "large red-dish-brown figure dash across the trail, from north to south." Higgins said, "The figure was upright and no smaller than Travis Lawrence, who is 6'3" and weighs approximately 250 pounds." Higgins ran to the spot where the animal had crashed into the woods, pistol drawn. He squatted down and caught a brief second glimpse of the creature as it strode through the trees to the east at a distance of only 20-25 yards. While watching the animal, he heard a "loud commotion from the northeast." Higgins thought this might be a diversion-ary tactic to draw his attention away from the retreating reddish-brown ani-mal. He kept focused on the original creature until it retreated out of sight. Higgins chose not to pursue the animal—as he only had a sidearm, a defen-sive weapon—and returned to camp.

The next event of significance experienced by Echo Team occurred at 3:25 a.m. on Monday, June 16. While on overwatch, Lawrence heard an "extremely loud sound." He said, "It was as if something had picked up and thrown a refrigerator onto the ground. It was the single loudest metallic sound I've ever heard in the valley." The huge noise failed to wake his sleeping teammates, however.

The next team to enter the valley was dubbed Foxtrot and consisted of Brian Brown, Walter Blake, Phil Burrows, and holdover Travis Lawrence. The team met up at the base cabin at 2:05 p.m. on Saturday, June 21. That afternoon and evening, the group would document wood-knocks, "heavy movement," and an odd "throat-clearing" sound.

After a relatively quiet Sunday, things picked up on Monday, June 23. Shortly after midnight, the cabin was pelted repeatedly with small rocks and/or nuts, making sleep difficult to come by. At 11:25 a.m., while eating break-fast on the front porch, Blake spotted a "large grayish-white" animal of some kind slinking through the brush south of the cabin in an east to west direc-tion. He could discern no legs, head, or tail, just a big "flat or horizontal back." Team members immediately hurried to various positions in an effort to flank or cut off the creature—unsuccessfully.

Between 6:00 p.m. and midnight, the team would document a strange "cough," multiple rock-throws, a series of "huffs," a canid-like "yip," a "deep-pitched guttural growl," and an "Ohio Howl-ish vocalization" an incredible 12 seconds in duration.

Just after lunch on the 25th, the men devised a scheme they hoped would yield a collection opportunity dubbed the "Haunted Mansion" plan. Everyone would pack up their vehicles and leave the compound. Only Lawrence would be left behind, hidden in the cabin's overwatch tower. The hope was that the locals would assume all the men were gone and walk out into the open.

At 8:01 p.m., the plan was set in motion. Lawrence's friends had been out of sight less than five minutes when he heard a rock slam the roof of the cabin. He wondered: *Are the apes probing to see if we are all really gone?* At 8:20 p.m., Lawrence heard "heavy movement getting closer" from the back side of the cabin. "I thought an ape was going to walk right up behind the cabin," he said. "I was prepared to fire through the window." But at 9:50 p.m., a hard rain began to fall, putting a damper on wildlife activity, and Lawrence retired to bed.

On the morning of Friday, June 27, Brown, Blake, and Burrows left for home. Lawrence stayed behind to await the arrival of the members of Golf Team. His wait would not be without some excitement.

At 8:50 p.m., Lawrence was startled when a gray fox jumped up on the front porch only six feet from where he was sitting. The fox's hackles were up and its back arched. "It looked like it was ready for a fight," Lawrence wrote. The small fox was gazing intently to the northwest. Seconds later, Lawrence heard the unmistakable sound of a large animal crashing through the woods at a high rate of speed. The animal was approaching from the northwest and was emitting "an awful low-pitched growl, accompanied by several snorts." The fox bolted off the porch and ran underneath Lawrence's truck. Lawrence then stood, shone a light to the northwest, began yelling, waved his arms, and unholstered his sidearm. The animal stopped, turned, and moved away to the west. Though he never saw the creature, Lawrence knew he had been bluff-charged. *I don't need this,* Lawrence thought and retired to the cabin.

Remember, Lawrence was alone. NAWAC protocols state that no member should attempt collection or take unnecessary risks when alone. We have since amended protocols and no longer allow individuals to stay in the valley alone.

At 9:45 p.m., Lawrence heard "an enormous crashing sound" no further than 50 yards from the cabin that was "like a large log being violently hurled upon the rocky earth."

The week of June 28 through July 5 proved to be a quiet one for Golf Team. While anomalous bangs, wood-knocks, rock-strikes, and movement were all documented, there was far less activity than previous teams had experienced.

The most significant event occurred on Thursday, July 3, at 12:30 p.m. Jordan Horstman had stepped away from his teammates near the fire ring in order to relieve himself when he spooked some sort of large animal in the wood line. Horstman was convinced the animal jumped down from a tree as he "felt it hit the ground and could feel the earth shaking as it ran off." The unseen animal crashed through the brush to the south toward the creek. "It was BIG and very heavy," Horstman said. The men investigated the area but found nothing.

Hotel Team, like Golf before them, found activity hard to come by during their week in the valley. Perhaps the most interesting event took place in the early morning hours of July 9 when Jeff Davidson, in his cot, awoke suddenly and heard "a large animal walking with heavy footsteps, of an even cadence—about one step per second." As the walker neared the tent, Davidson heard it growl. As Davidson attempted to wake the still sleeping Alton Higgins, the animal growled menacingly a second time. Meanwhile, Bill Coffman, sleeping in his truck in front of the main cabin and only 15 yards from the tent, was awakened by a deep growl along with several footsteps, as the walker, whatever it was, moved off to the north along the west side of his truck. Approximately 15 seconds later, all three men heard a loud "thud" from the east side of the cabin, as if a hand had slapped the wall of the structure. The "thud" woke up Gene Bass, who had been sleeping soundly near the east wall of the cabin.

By the afternoon of Saturday, July 12, the full contingent of India Team had arrived at base camp. Present were Travis Lawrence, Alex Diaz, Dave Cotter, and Paul Bowman. Their first two nights were quiet, but in the early morning hours of July 14 Cotter heard numerous loud rock-strikes followed by a "series of calls" he could not identify. It was "One long aaugh-oohaaugh," then a "shorter oohaaugh" followed by an even "briefer oohaugh." Those calls were followed in quick succession by six to seven calls of "ohah, ohah, ohah, ohah, ohah, ohah, ohah." An hour later, Cotter heard a "mumbling in the distance." That was followed a few minutes later by a sound reminiscent of "cracking, rotted wood." Cotter terminated overwatch at 6:30 a.m.

After a relatively quiet day, Cotter was standing in front of the east shed looking up the slope when he caught a brief glimpse of a "uniformly reddish-brown" animal moving east to west among some large, downed trees. "The color was similar to that of an Irish Setter," he said. "I clearly saw it...". Downed timber blocked a view anything but the top six inches of the animal's horizontal length. Cotter was adamant that he had not seen a fox, raccoon, deer, or bear (Cotter had seen many black bears in his home state of New Hampshire).

"I saw the front shoulders working under the fur or hair as well as the hips at the hind end...there was no tail," he said. He would add in his field journal, "I saw a severely sloping neck and/or back of the head."

At midnight on Thursday, July 17, the team retired to the cabin. Bowman and Diaz went to sleep while Cotter entered the overwatch tower. At 12:16 a.m., the cabin roof was struck loudly by a rock. Cotter scanned frantically with the thermal unit but saw nothing. At 1:10 a.m., the cabin was again struck by a rock. Again, the rock-thrower could not be detected. Five minutes later, the north-facing plastic "wall" of the overwatch tower was hit by a rock. At 3:45 a.m., Cotter heard a sharp impact on the wood frame of the overwatch structure near his left foot. The projectile struck the wooden frame and then bounced onto the roof of the cabin. Cotter was now certain that the tower was being purposely targeted.

Overcome by fatigue, Cotter climbed down from the tower, entered the main cabin, and walked toward his bunk. Just then, there were two "almost simultaneous impacts" on the wall of the cabin. The next morning, Alex Diaz would find "a rather large rock" on the roof of the cabin. It was the only evidence that anything unusual had taken place the night before.

India Team experienced only sporadic activity over the next two days. They left the valley on July 19 and were replaced by members of Juliet Team: Alton Higgins, Daryl Colyer, Phil Burrows, Ken Helmer, Jeff Henderson*, and his wife, Julie Henderson*. Colyer, in his role as team leader, announced his plans for the week: "I want to be very active and put constant pressure on the apes. I want to constantly probe, hunt, and play 'games' with them. My hope is that by applying constant stress on the apes one might eventually make a costly mistake."

That afternoon featured several wood knocks, a series of six low-volume growls, and two projectiles: one loudly struck the cabin roof, the other landed only ten feet from Colyer's position. Despite straining to see any sign of the thrower, Colyer could see nothing.

More wood-knocks were heard that night, as well as three metal bangs, which led Colyer to grab his .45-70/ATN thermal combo and glass down the trail leading back toward the west cabin. He immediately spotted a heat signature. It appeared small at first. "I have a hot signature at the end of the trail," he said. The words were barely out of his mouth when the signature suddenly became much larger. It appeared to be a large biped in a squatted position. Phil Burrows attempted to spot the animal by turning his powerful flashlight down the trail, but the animal, whatever it was, had already fled.

Around 2:30 a.m. on Sunday, July 20, Helmer, sleeping on a bunk in the main room of the cabin, was awakened by a sound on the edge of the roof just outside the window and above him. It sounded as if someone had dragged their knuckles across corrugated metal repeatedly. Moments later, he heard the something "shuffling" away on the hard ground outside the cabin. Perplexed, but very tired, Helmer went back to sleep.

After lunch, the team came up with a scheme they dubbed "the lost campers scenario." Three members would leave in their vehicles that night, the Hendersons would be cooking (a pork chop) dinner outdoors, while Colyer would hide out in the overwatch tower. The hope was that the resident apes would think the couple had been left alone at the cabin.

The plan was set in motion at 6:15 that evening. At 8:15 p.m., the Hendersons heard a very loud commotion up on the mountain slope above the cabin. Moments later, they heard a very large rock careening noisily down the slope, stopping only when it struck some scrap metal behind the cabin. The incident understandably unnerved the young couple.

At 9:47 p.m., two rocks struck the cabin in quick succession and a tree limb snapped loudly on the mountain slope. More unsettled than ever, the couple then heard something moving behind the wood shed no more than ten yards away. Jeff Henderson quickly glassed the spot with one of the group's handheld thermal units and saw a roundish heat signature near the back of the small structure.

"It appeared as though a head was peeking out from behind the shed," he said later. "The signature was very bright and clearly a living creature." Henderson watched the animal for a few seconds before it sank back into the forest to the west. Completely rattled now, Julie Henderson entered the cabin and woke Colyer, who was resting in preparation for overwatch that night.

"They're throwing big rocks and I'm really scared," Julie Henderson told Colyer. The Hendersons then retreated to the safety of their jeep, while Colyer entered the overwatch tower. Throughout the night, Colyer heard woodknocks, bangs on metal, and a couple of odd screams but never spotted his quarry. He aborted overwatch at first light. When Julie Henderson stepped out of the Jeep the next morning, she heard a rock clipping through vegetation. She looked up in time to see a softball-sized rock crash onto the top of the cabin before bouncing down practically at her feet. *That was thrown* at *me!* she thought.

The team hit the rack at 2:00 a.m. on the morning of July 22. Only Helmer remained awake in the overwatch tower. At 4:45 a.m., he heard the whistling of a sizeable stone flying over the canopy of the overwatch structure. The rock crashed to the ground roughly 20 feet to the south. Rattled, he contemplated the damage the stone could have done if it had come through the plastic walls of the overwatch tower and struck him.

NAWAC

Ken Helmer holds the large rock that sailed over his head while he was posted in the overwatch tower.

The shenanigans continued that morning and afternoon, and during the morning of the 23rd. Rock impacts, a rapping noise on the edge of the metal roof of the cabin, and the sound of footfalls were heard, but the team never caught sight of the perpetrators. And even on those rare occasions when the investigators did manage to spot the heat signature of an animal with a thermal scope, they were unable to make a positive identification.

A visual encounter did occur the afternoon of the 24th. Colyer and Higgins were in camp eating a late breakfast when they heard a "clear, musical wood-knock" ring out from the vicinity of the south cabin. Higgins grabbed his rifle and started east on the trail. Colyer began singing loudly in an effort to divert attention away from Higgins as he circled back toward the south cabin. A few minutes later, Colyer grabbed the binoculars and trained them on the south cabin area through whatever gaps in the vegetation he could find. After scouring the area for a few minutes, Colyer caught a brief glimpse of what appeared to be the head and shoulders of a wood ape through a small gap in the brush. Colyer sang out a warning to let Higgins know he was not alone. "It was grizzly bear brown," he said. "The area where the eyes should have been seemed lighter in color." They heard the breaking of sticks and rustling of brush from the west, followed by the sound of a deep wood-knock.

The men spent the rest of the day "beating the bushes" in an attempt to elicit contact with the locals but had no luck. A few minutes past 11:00 that night, the team was sitting around the dark fire circle when they heard a "raspy, roaring, multi-tonal" vocalization from the south. It was reminiscent of the Ohio Howl and, according to Higgins, "had a mournful quality to it." No more than ten seconds later, a second vocalization rang out from some-where south and west of the cabin. It seemed to be in response to the first call. Colyer and Higgins chose the same word to describe the vocalizations in their field journals: "creepy."

The following two days would be quiet by Area X standards. The team documented multiple wood-knocks, crashing trees, thudding footfalls, and a series of "huffs," but alas, there were no sightings or collection opportunities. Late Saturday morning, July 26, Colyer left for home. Higgins stayed behind and awaited the arrival of the members of Kilo Team.

Mark Porter and Bill Coffman arrived the afternoon of Saturday, July 26. The next two days were filled with the sounds of multiple wood-knocks and metallic bangs from all around the compound. Much of the activity seemed to be centered around the south cabin prompting Higgins to set up an obser-vation post in the middle woods. Just after midnight on the 29th, Higgins ob-served "multiple source heat signatures" moving quickly and in unison inside the wood line to his southwest. To Higgins, it appeared to be a large, upright animal. He realized that if the animal continued on its current course, he would have a collection opportunity and pulled back the hammer on the .45-70, which produced a loud "click." The animal immediately changed course and veered off into deeper woods away from Higgins's location. "It must have heard me cock the weapon," he said later.

Rain fell on the valley during the night and did not let up until Friday, August 1, seriously hindering the activities of the team. Coffman and Porter took advantage of a brief break in the weather to pack up and head for home. Once again, Higgins was left alone in the compound, but the apes declined to pay the solo investigator a visit.

Saturday, August 2, was quiet but for a "loud bang" Higgins heard from the south cabin area. "It is so odd to think there may be an ape 100 yards or less from me, and there's little to do about it," he wrote in his journal.

Higgins awoke with a start at 5:00 a.m. on Sunday, August 3, to the sound of someone walking in the cabin. "That is what it sounded like," Higgins wrote in his field journal, "the sound of walking in the cabin. I was instantly awake, listening to the creaking of the cabin in response to the heavy steps and the slow pace, I estimated at a bit less than one second between steps.

"For a moment I thought it must be one of our guys, but that didn't make sense because I had barricaded the front door with a chair. I lifted the curtain next to my cot and saw it was pitch black outside, just as it was in the cabin. The walking continued, and I realized it must be on the front porch. My best estimate as to how long the walking continued was about 30 seconds, then I heard it walk away to the east. I figured out which gun to take (I decided on the Lapua), turned on a dim red cap light, and the rifle's thermal scope, then walked into the front room. Everything was as I had left it the night before. I looked out the front window, but saw nothing in the dark. The visitor was gone from the cabin, but I thought it could still be in the vicinity, so I decided to enter the overwatch tower."

An admittedly unnerved Higgins conducted overwatch for the next two and a half hours but saw nothing and returned to bed at 7:40 a.m. Later that morning, Higgins added the following entry to his field journal: "I'm extremely frustrated at missing what may have been my best chance to collect a specimen. I just could not stay awake last night, constantly nodding off. No way I could have lasted another six hours for the arrival of the porch visitor.

"Thinking more about what happened, this person/creature may have stepped onto the porch like we do and walked up to the front door, which is actually two doors. Last night, the outer door was propped open with a rock, like it usually is during the day, but there is a sheet of thin trash bag plastic hanging in front of the inner door that is free to move in the breeze. I'm guessing this may have aroused its curiosity. I think it then backed away from the door and walked to the west end of the porch then walked back and left the porch the way it had entered. Did the early morning visitor leave when it realized that I had stopped snoring?"

In an effort to "semi-document" movement around the cabin, Higgins strung fishing line from the west wood shed to the southwest corner of the cabin at a height of six feet. He taped the cabin end so the line would pull out if something walked into it. He added a second fishing line trap 18 feet down the path that ran from his tent location to the south cabin. This line, too, was set right at six feet high. After setting his string traps, Higgins cleaned and swept the front porch thoroughly. "This morning's visitor probably won't return tonight, but if it does, hopefully, I will get some pristine prints as I'm going to sprinkle some of our hydrated lime [used in the outhouse] right where I think it is most likely to step on the porch," he wrote.

Higgins heard sporadic, odd sounds on and off the rest of the day but saw nothing. At 3:25 a.m. on Monday, August 4, Higgins was again awakened by a visitor near the cabin. "Creature returned!" he scrawled in large letters in his journal. Movement outside the cabin continued until about 4:15 a.m. when—after a wood-knock from the west woods—activity ceased.

In the morning Higgins chronicled the events in his notes: "I am sure the ape returned last night/this morning. It approached from the west, between the two sheds, making no attempt to be quiet, stepping on pieces of sheet metal at least twice. I could hear its footfalls and it appeared to depart to the south. It didn't sound as if it had gone to the porch, but I was hopeful that it had knocked down my string. Alas, the lime on the porch and the two strings remained undisturbed.

"My frustration is palpable. Could I be mistaken? What other animal can be heard walking with thudding footfalls on bare ground, or comes stomping around cabins in the wee hours of the morning for no apparent reason? Come to think of it, I heard the visitor fumbling around with something after it made its grand arrival. Maybe it was after a piece of firewood? That makes as much sense as anything else in this crazy place, because we've found several pieces of firewood in unexpected locations and have good reason to believe it is used to strike trees or rocks to create ringing wood-knock sounds."

Early Tuesday morning, August 5, Higgins was awakened from a deep sleep by the sound of something that seemed to be circling the cabin. He listened to the visitor as it shuffled about for the next 30 minutes, at which time Higgins quietly moved to the overwatch station in the front bedroom. Whatever had been pacing near and/or circling the cabin abruptly stopped and was not heard again for the rest of the night.

In the afternoon, Higgins gathered some jugs and headed west, past the compound gate, to get some washing water from the creek. Upon reaching the property gate, a movement caught his eye up on the mountain slope to his right. Higgins spotted an animal of "unusual color, a dark reddish-black mixture roughly the size of a raccoon." He could not discern a head or a tail. As

Higgins watched, the animal turned and slinked away in what was described only as "an unusual manner." Higgins dropped his water jugs and climbed up the slope in an effort to see the animal again but had no luck.

The wildlife biologist was perplexed as the color of the retreating creature did not match that of other animals of that size found in the valley with the possible exception of a bear cub. But the "unusual" gait of the animal had not seemed bear-like to Higgins. Could it have been a juvenile ape?

At 8:24 that evening Higgins heard "three long howls"—each about six seconds in length—from the west. The howls sounded "very close" and were deep in tone. Higgins described the vocalizations as "Ohio howl-ish" in his journal. Despite being a bit unsettled, the exhausted NAWAC chairman turned in at 8:45 p.m. Just he was beginning to doze off, he was startled by a "VERY LOUD VOCAL" that he described as being a cross between a "snarl, growl, and howl." He wrote in his journal: "More than anything, it was scary and very close to the cabin. I hope Travis [Lawrence] arrives soon." Higgins's wish was granted when Lawrence drove into the compound at 10:00 p.m.

August 6 featured a rock that struck the ground near Higgins. The rock-throw unnerved him as he had never had a stone thrown directly at him in more than a decade of research. Lawrence heard the distinct sound of "bipedal footfalls" moving in an easterly direction; however, he was unable to spot the walker. At 8:10 p.m. Higgins saw a "black, head-shaped object" moving through the wood line just west of the south cabin approximately 80 yards away. The visual was fleeting, but Higgins felt that he had seen an ape that had likely been searching for him.

Lawrence began overwatch on the 9th and at 3:00 a.m., the cabin was shaken by a powerful impact on the west wall. Ten minutes later, "an absurdly loud wood-knock" erupted from the southeast. The sound was so powerful, Higgins—awakened from a deep sleep—thought Lawrence had fired his weapon. "It was the loudest wood-knock I've ever heard," Lawrence said. At 9:20 that morning, both men heard a "BIG, explosive, loud, cracking-smashing-to-the-ground-tree fall" from the south. Higgins commented, "It sounded close."

After spending three straight weeks in the valley, Higgins departed that afternoon. Lawrence stayed behind and awaited the arrival of the members of Mike Team, Andrew Hein, Dave Cotter, and Gene Bass being the first members to arrive. The valley was quiet over the next 36 hours save for the occasional sound of a rock striking one of the compounds satellite structures. That would change on the afternoon of August 11.

After a morning filled with the sounds of wood-knocks, most of which emanated from the east, Lawrence and Bass decided to strike out in that direction. At 3:15 p.m., the pair was roughly 75 yards south of the east cabin when Lawrence heard a soft "huff." He immediately turned and heard the sound of an animal sliding down a tree and thumping on the ground. Lawrence then saw a "small, dark-colored animal" moving through the underbrush.

"Look at that!" Lawrence yelled to Bass. There was more than one animal moving through the brush. The men witnessed "four, small, hair-covered animals that looked just like small chimps" move quadrupedally away from them. Lawrence noted that the animals moved in a single-file manner as they made their getaway. "They were small; no more than 30-40 pounds and about two feet high," Lawrence said. "Their upper bodies were longer and wider than their lower bodies and they seemed to move with their heads down."

He stressed again how similar their appearance and locomotion were to that of chimpanzees. (Chimpanzees are not native to Oklahoma. The only chimpanzees in the state are found in zoos or primate rescue facilities far from the Ouachitas.) The color of the animals ranged from "reddish-black" to "reddish-brown." The two attempted to follow the animals but gave up the pursuit after hacking through the bush for 500 yards. Lawrence believes the subtle "huff" he had heard had likely come from an adult ape that was directing the four young animals in the tree to leave the area. Stunned, the men returned to camp.

Investigators Jerry Hestand and Mark Porter arrived at the compound on August 13. While the new arrivals unpacked, the other investigators made plans for a night hunt. The scheme was for Bass and Cotter to approach the south cabin from two different directions while carrying flashlights. Hein, who had donned a ghillie suit, would approach the same cabin from a third direction and conceal himself near the northeast corner of the structure where the porch meets the overgrown and fenced-in "yard." Once deployed, Hein began to hear animal movement near his location almost immediately. The most worrisome sound was the displacement of rocks in the dry creek bed just on the other side of the cabin. Still, he held his position. Just before 10:00 p.m., Hein tilted his head back to stretch and admire the multitude of stars in the clear sky overhead. This small movement seemed to trigger the local wildlife, and Hein heard movement on all sides of the cabin. The investigator froze, not wanting to move again and give away his position. Hein felt that it was very possible an ape could step out in front of his location at any moment. As these thoughts crossed his mind, he heard a heavy "thud" on the porch just behind him. Moments later, he felt a gentle pressure on his head-lamp (which was off) and hat "as if someone had reached out and touched the light." Hein yelled, "Shit!" and sprang from his position, thrashing through vegetation as he went. He clicked on his headlamp and raised his rifle but could see nothing through the thermal scope. Unnerved, the investigator "beat a hasty retreat back to the base cabin."

Porter was initially skeptical of his teammates account and said that Hein should have heard any animal moving through the overgrown "yard" area before it could reach him. Hein could not discount Porter's logic but stuck to his story. The next day, the group visited the south cabin and arrived at a hypothesis. If an ape had been concealing itself in the fenced-in "yard" area of the cabin, it would have been very easy for it to approach Hein from behind.

The remainder of Mike Team's stay in the valley was relatively quiet, and the investigators departed for home on Saturday, August 16.

The valley would not stay vacant long as the members of November Team arrived at 3:00 that afternoon. The team consisted of Rick Hayes, Phil Burrows, Laura Altom, Hannah Altom, and holdovers Hestand and Porter. It did not take long for things to get interesting.

At 6:40 p.m., while returning from a short hike to the compound gate, Laura Altom saw a flash of movement in the treetops. She described seeing an animal "dark in color and weighing perhaps 30 pounds." She watched as the animal leaped from one limb to another. "It looked like a monkey," she said. Further investigation yielded no results, and the team returned to the base cabin.

On August 17, Laura Altom deployed a "pink blow-up kiddie pool" in the bottleneck area east of the cabin. She added a "crying baby doll and stuffed Chewbacca doll" to the scene and hung "metallic pom-poms and brightly-colored feather boas" in the brush. She hoped the unusual objects would pique the interest of an ape. It may have as the team heard two "howls from an unidentified animal" shortly before 9:00 that evening followed by odd sounds and wood knocks over the remaining evening hours.

At 1:07 on the morning of the 18th, Hayes and Hannah Altom, while conducting overwatch, heard "rustling" followed by clear "bipedal footsteps" very near their location in the tower. Hayes was so sure a shot opportunity was about to present itself that he whispered, "When I flip the safety, cover

173

your ears," to his partner. Meanwhile, down below, Burrows also heard the footfalls and was "so alarmed" by this encroachment that he turned to warn Hayes. Inexplicably, at that moment, the cabin's water pump kicked on and whatever had been walking near the cabin retreated.

At 7:52 a.m., Laura Altom reported hearing what "sounded like a car door slamming" from the south cabin area. Veteran researchers Burrows and Hayes, who were familiar with this sound, immediately suspected one of the dead freezer lids had been slammed. When the group investigated, however, they found rocks on the lids of the freezers. Puzzled, the group returned to camp. The remainder of the day was filled with the sounds of rock-on-metal, anomalous bangs, and another "car door slam." This slamming sound greatly troubled the Altoms, who had not heard it before, as they knew they would have heard or seen any vehicle coming into the compound. While I wasn't there, I suspect strongly it was the same sound so often attributed to the slamming of the dead freezers.

November Team exited the valley on Saturday, the 23rd. A week would pass before NAWAC personnel would return to Area X as the property owner and his family were staying in the cabin. Papa Team—made up of Bob Strain, Kathy Strain, Walter Blake, and Ken Stewart—arrived on Friday, August 29. When they arrived, the owner of the west cabin was present and informed the team that he had found "two new rocks" on the metal roof covering the west RV. He also reported multiple "new dents" in the side of the RV and said he was going to "start marking the dents so he could tell new ones from old ones."

At 12:40 a.m. on the 31st, Blake reported hearing something strike the cabin roof. Just 20 minutes later, the northwest corner of the cabin was struck with a "good degree of force." Kathy Strain described the sound as "an extremely hard...cabin slap." The rest of the night would prove quiet. Blake and Stewart would exit the compound the next morning. The west cabin owners also took their leave, leaving the Strains alone.

Things remained fairly quiet with only a few rock-strikes and light wood-knocks documented until 8:01 p.m. on Tuesday the 2nd. At that time, a pack of coyotes began yipping and howling back to the east. When they stopped, the Strains heard "two strange animals...calling in very high-pitched screeches" from up on the mountain slope. "Birds?" Kathy Strain wondered in her field notes. After conducting overwatch until 1:00 a.m. on the 3rd, the couple went to bed.

At 5:30 a.m. "something hit or shoved the back wall of the cabin near the shower area," Kathy Strain wrote in her journal. "It sounded like another slap." Investigation yielded nothing. Three hours later, "a very strange sound" woke the Strains. "It sounded like someone was moving about inside the cabin," said Kathy Strain. "Then we heard the sound of an animal shaking water off...like a dog." Bob Strain grabbed his rifle and cautiously exited the bedroom to inspect the cabin. No one was there.

Kathy Strain was awakened on Friday the 5th by the sounds of rocks slamming the roof and back wall of the cabin. The activity would continue with multiple wood-knocks ringing out from the south and west over the next hour. At 9:33 a.m., the couple heard "what sounded like someone talking in the middle woods," wrote Kathy Strain. Investigation yielded nothing.

At 5:55 that afternoon, just before Ken Stewart drove into camp, a "large rock" banged loudly off the south cabin. Had the rock-strike on the south cabin been some sort of alert? In any case, the locals seemed to be feeling their oats as a rock barrage began at 8:10 that evening and continued until well after midnight. The assault resumed at 7:38 the next morning. While not continuous, rocks continued to fly all over the compound, slamming the walls and roofs of the various cabins and sheds located there.

The Papa Team holdovers officially became members of Quebec Team upon the arrival of Brian Brown Saturday afternoon. Just before 9:00 that evening, a now familiar sounding "Whumpf!" sound erupted from the area of the south cabin. Everyone immediately suspected the lid of one of the dead freezers had been slammed. When they rushed over to investigate, they found the previously placed rocks still on top of the freezer lids. The team was baffled. They were certain the sound they had heard had been a freezer lid slamming. How could the rocks all still be in place? Could an ape have picked up one of the rocks, slammed the lid, and put the rock back before fleeing? "I do not enjoy the thought of that possibility," said Bob Strain.

At 10:30 p.m., Brown started broadcasting agitated chimpanzee vocalizations. An hour later, all heard a "strange vocalization" from the middle woods, followed moments later by the sound of "muffled and garbled human-like speech." The team noted in their field journals that "It sounded like human females talking."

Monday, September 8, dawned brightly after a second consecutive quiet night. The first rock did not strike the cabin until 7:05 a.m. Two hours later, a projectile blasted the south cabin. Kathy Strain, who was standing on the porch, caught a glimpse of a large gray animal moving through the foliage. She saw the animal from the neck down as it quickly darted across a small gap in the brush before disappearing into thicker cover. The men rushed over to the spot of the visual. After determining the animal was no longer in the area, the men recreated the sighting. Kathy Strain was able to discern that the animal she had seen was easily twice the width of her husband, Bob, and another foot taller (this would have made the creature at least seven feet tall). The rest of the day was filled with the sounds of rocks raining down on the various cabins and the west RV.

New Zealander, Daniel Falconer, who would be acting as a non-NAWAC affiliated independent observer over the next week, arrived at 2:00 p.m. on the 9th. The sounds of rock-strikes and wood-knocks continued periodically during the morning and mid-afternoon hours. At 7:26 that evening, Brown and Falconer were at the main trail just northwest of the RV, when they heard a clear wood-knock emanate from the south near the creek. As they turned to look in that direction, Brown caught a "quick flash of large movement" 50-60 yards distant. He focused his attention on the area and was rewarded with the sight of a large, dark, upright figure "halfway emerged from behind a tree," seemingly peering at the men. Brown said, "Oh, shit!" and called out to Colyer, who was back in the woods scanning the mountain slope. Falconer, who had caught a glimpse of movement himself, added, "We have an animal!" Colyer began crashing through the foliage in an effort to reach his teammates but the thick brush impeded his progress. It took him nearly two minutes to cover only 100-150 yards then continued south in search of the creature. Hearing the animal crashing through the brush and increasing the distance between them, Colyer realized further pursuit would be folly, and the men returned to camp.

Friday, September 12, was scarcely 45 minutes old when a rock blasted the roof of the south cabin. Ten minutes later, Brown and Colyer both heard what they described as "undecipherable human male talking" from the southwest. At 1:08 a.m., the team was startled by the sound of falling water coming from a tall tree overhanging the east shed. "The sound," Falconer wrote, "evoked an impression of a nimble animal either shimmying up or skidding down the tree, setting off a cascade of water from the sodden foliage." When they reached the tree, which was still shaking, there was no animal about.

The men in the group took a hike toward the property gate at 3:30 the next afternoon. At one point, Colyer, seeing an odd shaped tree in the distance, asked for the binoculars Brown was carrying. As he did so, the "tree" moved slightly. It was at this point that Bob Strain observed what he described as a "honey-colored head" rise above the deadfall briefly before disappearing behind it. At that moment, a wood-knock erupted and Colyer, Falconer, and Brown immediately sprinted into the forest toward the animal. Bob Strain remained in place in case the animal somehow doubled back. While searching the area, a rock landed among the three investigators, seemingly thrown from the south. The men could only shake their heads.

Later than afternoon, while Colyer was making his way back toward camp, he spotted the honey-colored, dome-shaped head of an ape clearly visible in a window of vegetation near the west cabin. The animal appeared to be looking directly at him. Colyer began slowly raising his rifle, but the head ducked away before he could take a shot. Colyer sighed and returned to the base cabin.

Determined to give it one more try, the team returned to the area just south of the west cabin. About 7:20 p.m., Colyer, in reaction to the quickly fading sunlight, decided to "dial down" the size of the red dot projected by his Aimpoint CompM3 scope. After clicking the scope, he looked to his southwest and observed a "large brown mass" the approximate size of a door frame step into a narrow opening in the trees 60 yards away. *Holy shit!* Colyer thought and clicked his rifle off safe. As he began to raise his rifle, the animal bolted to the north and out of sight.

"It must have seen my movement," Colyer later mused. "Damn it!" he cursed in frustration. The words had barely left his mouth before he saw a second creature streak across the same window in the vegetation where the first animal had been standing. Colyer immediately bolted and tried to intercept the apes at the main trail to the north. But once there, he saw nothing. Then an explosive wood-knock rang out from the area he had been hiding for the last couple of hours. Stupefied, and more than a bit unnerved, Colyer shuffled back to camp.

After a quiet night, Falconer was awakened at 6:50 a.m. on the morning of the 13th by the sound of "mumbling" outside the west wall of the cabin. The Strains left for home later that morning. At 10:45 p.m., no more than five minutes after turning off the lights, the remaining team members heard the sound of running footfalls along the west side of the building, followed by a light impact of some kind on the west end of the porch. Before the men could move to investigate, the footfalls trailed off into the night.

The last likely ape-interaction of 2014 was a comical one. At 3:15 on the morning of Sunday, September 14, an object of some kind struck the back wall of the cabin near the spot where an exhausted Daryl Colyer was bunked. "Knock that shit off!" he yelled. He was answered by a "huff." Part of me wonders if it was a laugh.

The team was up early and left the valley at 9:00 a.m., bringing Operation Tenacity to a close.

18
Operation Resolute

The NAWAC initiated its fifth long-term field study in Area X on May 30, 2015, with the arrival of Albatross Team: Rick Hayes, Tony Schmidt, Andrew Hein, and David Haring. The men reported a few wood-knocks, small impact sounds, and a "loud commotion" in the middle woods over their first 36 hours in the compound. The only wildlife directly observed was a doe and her fawn that seemed to be sticking close to the base cabin for reasons unknown to the team.

The most exciting event of the first two days was a surprise encounter between Schmidt, Hayes, and a black bear at 9:25 p.m. on the 30th. The bear seemed just as startled as the men and quickly fled. But the encounter would foreshadow some bear problems that arose over the course of the summer.

The next evening, Schmidt decided to hunt from a tree stand in the woods west of the base cabin. At 9:30 p.m., he requested assistance from his teammates as the climbing stand he was utilizing had proven to be highly unstable. After helping extricate Schmidt, Hein scanned the area with a handheld thermal unit and spotted a "large thermal image big enough to be a bear or an ape" to the northwest. When the men attempted to light up the animal, they saw "blue-white eyeshine." Attempts to flush the animal failed, and the men returned to camp.

At 10:30 on the night of the 4th, the entire team hiked east. Schmidt quietly left the group upon arriving in an area just north of the east cabin. The remainder of the group—red headlamps on—continued their hike. Roughly one hour later, while scanning with the thermal site on his rifle, Schmidt spotted a bright white heat signature to the west-northwest. The figure was low to the ground and partially obscured by a large pyramid-shaped boulder. As he watched, the "tango" moved slightly downward, as if trying to better conceal itself behind the rock. Schmidt quietly reported the visual to his teammates via radio who assured him they were not in the area.

Schmidt, now firmly convinced he was watching an ape, prepared to fire. He rested the barrel of his rifle on the top of his left hand, which still held the walkie-talkie, and sent the round. Schmidt dropped the radio, re-gripped the weapon, chambered another round, and attempted to reacquire the animal with the thermal scope. There was no sign of it.

Upon hearing the shot, the rest of the team activated white lights and rushed to Schmidt's location where they immediately began a search of the area. The team searched for approximately ten minutes without success. Hayes then advised the others that "backing away" from the scene might be advisable. If the animal was wounded, he reasoned, aggressive searching might drive it farther into the bush. Despite an extensive search later that night and into the next day, no sign of the animal Schmidt targeted was ever located.

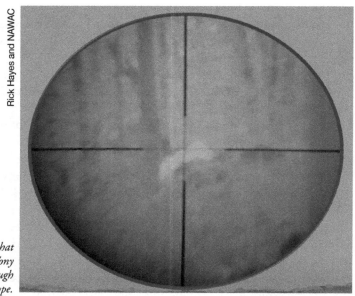

Rick Hayes and NAWAC

A recreation of what investigator Tony Schmidt saw through his thermal scope.

Blackhawk Team was made up of Travis Lawrence, Jordan Horstman, Jerry Hestand, Marvin Leeper, Mark Porter, and Bill Coffman. The team was fully assembled by the night of the 6th. The only wildlife spotted that evening was a black bear moving east of the cabin and an exceptionally large raccoon that helped itself to some cereal the men had left on the front porch.

The night of Monday, June 8, had been quiet, so Hestand, Coffman, and Leeper decided to do some "road-running" in Hestand's jeep in an effort to stir up some activity. At 10:00 p.m., after turning onto a crude side road more than a mile from the property gate, the men spotted a "large gray animal" quickly drop to all fours and disappear into a tunnel of vegetation. Hestand noted, "The animal was very large and gray, almost white, and had huge legs and buttocks." The men quickly exited the jeep and sprinted to the opening in the thick brush. Leeper entered the "tunnel" and had traveled about ten feet when he caught a glimpse of two green, reflective eyes. The creature then turned away and retreated out of sight.

"It was an ape," Leper later wrote, "not a bear. The animal's hindquarters were more visible than the shoulder or head. The head, face, etc. appeared round, like a basketball, with no ears or snout visible. The shoulders were a bit lower than the head and at least four feet from the ground. The spine was not straight but slightly sway-backed. The hips and buttocks were massive. The hind legs were huge…the animal had an odd posture and leaped over a lot of ground in no more than four steps. It was leaner than a bear but massive."

At 8:00 p.m. on Thursday, the 11th, the men returned to the site of the visual. Hestand retrieved two gray hairs from the "tunnel" and bagged them for future forensic analysis. The remainder of the week was filled with the usual rock-strikes and wood-knocks, but no other visuals occurred, and the members of Team Blackhawk, with the exception of Travis Lawrence, left the valley on Saturday, June 13.

Condor Team consisted of Alton Higgins, Ed Harrison, Justin Horn, Dusty Haithcoat, and Travis Lawrence. Higgins set up his tent and deployed a motion detector on a tree behind the cabin. The motion detector was encased in a plastic shell that looked like a Northern Cardinal and made chirping sounds when it sensed movement. The "bird" was placed on a tree on the northeast corner of the cabin about six feet off the ground.

That evening and the next day were punctuated by the occasional high-pitched wails, knocks, and bangs. But at 2:30 that afternoon, Higgins was sitting on the east side of the cabin when he saw an unusual-looking stump that he did not recall seeing before. After observing the stump for several minutes, Higgins started down the trail to check it out. As Higgins hiked east, Lawrence—who had stayed behind—caught a flash of movement on the mountain slope about 70 yards above Higgins. Though he saw it only briefly, the animal had been white or gray. Lawrence was sure he had not seen the tail of a deer. Later, Higgins heard "three high-pitched screams" from the west. Overwatch that night yielded no visuals, but the cabin was struck multiple times by projectiles presumed to be rocks.

At noon on the 15th, the men checked the cabin roof and located three small rocks. The rocks had to have been recent additions as Jerry Hestand had swept the roof the previous week. At 2:00 p.m., Higgins decided to set up some string traps in the woods west of the cabin. He was just inside the brush line, only 15 yards from the cabin, when a rock whistled uncomfortably close by. Higgins believed he had been targeted. Before he could react, he heard a "humph wah" vocalization to his south. Higgins and Lawrence spent the next hour attempting to locate the rock thrower without success.

The next significant event occurred at 5:05 a.m. on the 18th, when Harrison, who was conducting overwatch, observed a faint gray heat signature about the size of a dinner plate in the brush just south of Higgins's tent. The signature was round and became whiter as the animal, whatever it was, moved closer to the tent. Harrison watched for approximately three minutes but was unsure what he was seeing. Eventually, the animal moved off into the brush. The next morning, the men recreated the event and discovered that what Harrison had seen had to have been at least six to seven feet off the ground. Harrison grimaced in disgust, realizing he had likely been watching an ape. His teammates reassured him that he had done the right thing by not firing on an unidentified target.

Harrison was conducting overwatch when Lawrence and Horn, who were sitting on the front porch, detected a "very musky animal smell" at 12:30 a.m. on the 19th. The pair decided to retreat inside the cabin in the hopes that their absence would tempt the animal into approaching closer and into the view of Harrison. An hour later, Harrison felt a "soft bump" on the northwest corner of the cabin. Horn, sleeping on a cot in the northernmost room, was awakened by the bump and heard movement outside. Seconds later, the faux bird motion detector began chirping, but nothing was ever seen.

Darter Team relieved Condor on Saturday, June 20. The team consisted of holdover Alton Higgins along with Phil Burrows, Dusty Haithcoat, and Ken Helmer. That night, Helmer decided to play agitated chimpanzee vocalizations on his game caller. Within 17 minutes, loud banging was heard from the east. As Saturday night yielded to Sunday morning, more anomalous banging was heard from the west cabin area. The remainder of the nighttime hours would be intermittently interrupted by rock-throws, banging, and wood-knocks.

Helmer and Haithcoat left for home the next day, leaving only Higgins and Burrows in the valley. The two remaining investigators spent the afternoon deploying new string traps near the west cabin and on the north side of the bottleneck trail.

While the men were eating a late breakfast on the morning of the 23ʳᵈ, the motion-detecting cardinal behind the cabin began to chirp. By the time the men arrived, whatever had set the motion detector off was gone. But a large spider web that had stretched between the tree and the northeast corner of the cabin had been obliterated; Higgins had noticed the ornate web and photographed it the day before. "Well," Higgins sighed, "that spider web didn't disappear on its own; something knocked it down."

At 2:50 in the morning of June 24, Higgins, who was sleeping in his tent, was awakened by the sound of something moving through camp. He heard items being tampered with and moved about and a sound he took to be a large animal stepping upon onto and then off the front porch of the cabin. The previous evening, a large raccoon had been caught rummaging through the supplies located on the front porch, and Higgins figured another masked bandit had decided to help itself to their food stores. Higgins clapped his hands loudly in an effort to spook the critter and get it to retreat.

After a few moments of silence, Higgins heard the creature begin to move away. When he heard the heavy lumbering footfalls, which were spaced evenly about one second apart, he realized an ape might have entered camp. Higgins immediately grabbed his flashlight and exited his tent. The animal was nowhere in sight, but he heard the sound of crunching movement to the west. Higgins then heard a "human-like coughing" sound followed soon after by a loud bang from the west RV area.

At 4:00 a.m., the creature returned, though it stayed in the wood line and out of the field of view of Burrows in the overwatch tower. Both investigators heard a "very angry-sounding roar or growl erupt from the previously silent west woods." The vocalization lasted four seconds and was extremely intimidating. Between 4:15 and 4:44 a.m., the men heard loud bangs, rock-strikes on the cabin, and wood-knocks from the west.

Higgins and Burrows left the camp to re-supply early the next morning. Upon their return that afternoon, they found that a black bear had ransacked their gear and supplies. Higgins was especially distraught, as the bear had eaten or destroyed supplies he had intended to last nine more days.

The incident caused the men to reconsider what their nocturnal visitor from the night before might have been; however, upon reflection, the pair decided their nocturnal visitor had likely not been a black bear for various reasons. First, black bears are diurnal and not usually active at night. Second, the supplies had not been ransacked, bitten, or torn up in any way. Third, the sound of the creature's footfalls after Higgins clapped his hands suggested a

biped, not a quadrupedal animal. Finally, the nocturnal visit had been followed by anomalous bangs and a threatening growling/roar that Higgins, a
wildlife biologist, knew had not come from a bear. The difference in behavior
between the larcenous bear and the mysterious night visitor was stark and, in
Higgins's mind, clearly pointed to two different culprits.

On Friday, June 26, the two men were awakened at 2:00 a.m. by a
"rain of rocks" that pelted the cabin "fairly constantly" for an hour straight.
(The "Rain of Rocks" audio file can be heard on the book's webpage at
anomalistbooks.com.) The investigators were unable to see anything in the
surrounding darkness.

Investigator Angelo Landrum joined the team on Saturday, June 28.
During a conversation, a pack of coyotes began howling back to the west.
Higgins noted in his journal that these were the first coyote vocalizations he
had heard in weeks and that it was "odd to hear the howling in broad daylight."

The next morning, Higgins and Landrum walked to the west cabin area
to check on the string traps. The string on the west side of the RV had been
disturbed. Initially, the string had been stretched north-to-south at a height
of roughly five feet five inches across what looked to be an east-to-west running game trail. Something had walked through the string while traveling to
the northwest. Higgins noted the disturbance and pondered on whether this
location might be a good spot for the deployment of one of the small nano
tracker radio tags the group had recently purchased.

By Tuesday, the 30th, investigators Travis Lawrence, Shannon Mason, and
Shannon Graham had arrived in camp. Upon their arrival, Higgins suggested
that Graham and Mason hike all around the compound in order to "adver-
tise" the arrival of two females. At 2:20 that afternoon, the team heard "three
loud slams" coming from the south cabin, likely the dead freezers lids being
slammed. The team hustled over to investigate, but upon arrival, they found
that the rocks previously placed on top of the freezers remained in place.
Puzzled, the group returned to the base cabin. They would exit the valley on
Friday, July 3.

On Saturday July 11, the Falcon Team arrived: Marvin Leeper, Gene Bass,
Mark Porter, Jerry Hestand, Chad Dorris, Walter Blake, and Blake's daughter,
Sarah. Two days later at 4:00 a.m. a fox was heard whining behind the cabin.
The team moved to investigate and found the small fox running back and
forth, obviously quite distraught, near the base of the mountain. Suddenly,

they heard the sound of a large animal crashing through vegetation on the mountain slope. Whatever it was, it was big and moving "very fast," but thermal scans yielded nothing.

Per Leeper's request, Sarah Blake provided a urine sample to be placed in a "drip bag" in the forest near the swimming hole east of the compound. A game camera was deployed overlooking the bag. Leeper hoped that the scent of human female pheromones would be too much for the locals to resist and an ape would be lured in front of the camera. The men checked and replaced the drip bag several times throughout the week, but acquired no photos.

At 1:00 a.m. on the 16th, one of the more frightening events in NAWAC history took place, but it had nothing to do with an ape. Most of the team had retired for the night, but Chad Dorris was not quite ready to go to sleep and, instead, decided to take a hike. On his way back to the compound, Dorris paused at the final creek crossing before entering the property gate and scanned upstream with a handheld thermal unit. Doing so, he saw a large cat parallel to him at a distance of no more than ten yards. When the cat turned toward him and crouched low to the level of the rocky creek bed, Dorris realized he had a serious problem. Dorris dropped the thermal, raised his Remington 870 to the low ready position, and snapped on the weapon's tactical light. By the time he reacquired the now illuminated cougar, it had crept to within two yards from him. No longer having any doubt as to what the puma's intentions were, Dorris fired his shotgun—loaded with 00 buckshot—into the ground at the big cat's feet. The long-tailed cat turned and fled and a rattled Dorris double-timed it back to the safety of the base cabin.

The next morning, the team decided there would be no more solo hikes or remote campsites for the remainder of the week. There was some disagreement about whether or not Dorris should have dispatched the cougar. In Dorris's mind, killing the cat was simply not necessary; other teammates felt that having a mountain lion in their vicinity was a very dangerous situation. But they all finally agreed that, should a similar situation occur in the future, the big cat would be shot on sight. No further encounters occurred, however, and all team members left the valley on Saturday, July 18.

I, along with Gull Team members Alton Higgins and Phil Burrows, arrived in camp at 5:40 p.m. that day. Upon arrival, Higgins began the process of checking his various string traps. He quickly discovered that something had broken through a string set six feet high behind the base cabin. Later, while working on the Marlin .45-70 overwatch rifle (the weapon had jammed and we were attempting to extricate a round from the chamber), Higgins and I heard some kind of faux speech or chattering from behind the cabin. These sounds were quickly followed by a "hwah" vocalization. Investigation yielded nothing.

The next morning, while hunting on the mountain slope, I heard a vehicle enter the compound. The driver was honking his horn loudly. As we were not expecting anyone, I went to discover the identity of our visitor. I was first greeted by a friendly black lab wearing some sort of radio collar. I then saw an unfamiliar truck pulling into camp. A bald man exited the truck and called to the dog. The man seemed a bit taken aback by my appearance—I was heavily camouflaged and carrying a semi-auto AR-10—but explained he was searching for his dogs that had been "missing since yesterday." The man gathered his black lab, got in his truck, and left. I remain unconvinced that he was telling the truth; after all, he had ignored multiple warning signs and opened a closed gate to enter the property. Shortly after the gentleman's truck was out of sight, a rock loudly struck the cabin roof. I gave the unseen rock-thrower a hand gesture that I think is universally understood across the primate world and retreated to the front porch.

Ken Helmer arrived on Tuesday, July 21, and the team drove into town and dropped the jammed .45-70 off at a gun shop for repair. To our surprise, the gunsmith was able to quickly restore the gun to a working condition. He did ask, "What are you guys hunting up there, elephants?" Helmer smiled and replied, "Something like that."

Higgins and I began scouting for locations in which to place the NAWAC's new nano tag radio trackers, which were now ready for deployment. The small tags would be placed inside hollowed-out cockleburs and hung on various string traps of Higgins's design. We hoped that when an ape walked through the string, the cocklebur/tag would become hopelessly entangled in the creature's hair, allowing it to be tracked. During this scouting operation, we found that strings near the bottleneck area and behind the cabin had been disturbed. In all, the string traps had been compromised five times during Operation Resolute, inspiring high hopes that the radio tag experiment would yield fruit. Investigator Ken Stewart arrived with the nano tags late Friday night. Phil Burrows and I would leave the next morning, but Higgins and Helmer stayed one more day to get the nano tags deployed and ensure that the tracking system was fully operational. All of the radio-tags had been successfully deployed by the afternoon of Saturday, the 25th. It was now a waiting game.

Just after midnight on the 26th, the team heard some kind of faux speech or moaning behind the cabin. Twenty minutes later, the team discovered that one of the tags in the bottleneck area had been activated. But it turned out to be a "false activation." No cause for the malfunction was immediately obvious. Higgins and Helmer left the valley the next morning.

Paul Bowman and Angelo Landrum were now the only members of Harrier Team on site. Over the next two days, the men documented rock-throws, mouth pops, a "loud, quick whistle reminiscent of a Bobwhite Quail's call," two high-pitched yells, and an unnerving "enormous sound" to the west of the cabin described

NAWAC

A nano tag ready for deployment.

as "sounding like someone lobbed a Volkswagen." It was clear to all that no animal known to inhabit the valley—including black bears—was capable of creating such a huge sound.

By the early afternoon of Saturday, August 1, the Ibis Team—Mark McClurkan, Robert Taylor, and Gene Bass—arrived at the compound.

At 6:00 p.m. on Monday, the 3rd, Taylor hiked to an area south of the southern-most cabin and requested that McClurkan produce a series of whoops in the hopes of distracting any apes lingering about the area. Five minutes later, McClurkan observed an "upright figure" moving east-to-west on what appeared to be the main trail, but he only caught fleeting glimpses of the figure. He immediately radioed Taylor and asked if he had been in the area. "Negative," replied Taylor.

Three days later, McClurkan left camp to pick up investigator Rick Mc-Daniels*, who did not have a vehicle capable of negotiating the rough road into the valley. On the way back, both men spotted a large, upright animal moving "at an incredibly fast clip" north to south across a small clearing in the woods. Both men immediately recognized the creature as a brown ape. "It had broad, muscular shoulders and ran in a somewhat crouched position as it seemed to be in a full sprint," McClurkan said. "It moved smoothly, with little to no bounding in its gait." Stunned, the men continued into camp.

McClurkan, Bass, and Taylor left the valley on Saturday, August 8, ceding control of the camp to Brian Brown, wildlife biologist Lloyd Merchant, and holdover Rick McDaniels. A unique and somewhat confusing sighting event took place in the wee hours of Monday, August 10. The team was sitting in front of the cabin at 1:15 a.m. when McDaniels spotted "something with a conical head and shoulders" crouching low to the ground about 15 yards east of the outhouse in the middle woods. McDaniels notified Brown and handed him the thermal unit. For some reason, Brown looked toward the mountain slope instead of the area behind the outhouse. He immediately observed a "huge, crystal-clear silhouette of a classic wood ape shape…pointy head broad shoulders, BIG!" McDaniels and Brown passed the thermal unit back and forth multiple times, arguing about where the ape in question was located. When Brown got up to approach the animal on the slope, an agitated McDaniels, who was tailing Brown, did not understand why they were going toward the mountain and not the middle woods, where he had seen the creature. Brown lit up the slope with white light and saw a "flash of gray movement" before the animal disappeared.

It was only after returning to camp and doing a recreation of the event that the men were able to figure out that they had been observing two different heat signatures in different locations. "Rick went up the slope and stood in the spot…" Brown wrote later. "I realized the window in the foliage to my position was perfect and exactly where you'd stand if you wanted to see what a bunch of weirdos were doing sitting in a big circle only illuminated by starlight. Finally, I was able to understand exactly how BIG that animal was. Every bit of eight feet tall and at least five feet wide."

On Thursday, the 13th, Brown heard a "ridiculously loud impact" from somewhere to the west. Later that night, one of the camp foxes wandered into camp "acting strange and wary." The team could not discern if it was sick or afraid. At 10:44 p.m., the men heard a vocalization that sounded "every bit like a man screaming" in the west woods. Investigation yielded nothing. The team packed up and left the valley at 9:00 a.m. the next day.

Most of the members of Kingfisher Team—made up of Gene Bass, Dave Cotter, Scott Wheatley, Dusty Haithcoat, Alton Higgins, and Mark Porter—rolled into the compound on Saturday, August 22. The men inspected each of the nano tags and string traps and discovered that tag 7, located on the trail that ran through the woods between the south cabin and north cabin, was down. It was unclear on whether an animal had walked through the string or if weather conditions had dislodged the tag. The men reset the trap. Just after lunch the next day, Cotter found that the tag 7 trap had, once again, been disturbed. Cotter began to suspect, due to the trap being compromised twice in a small window of time, that an animal was using the trail and walking through the trap. If so, he thought, it was only a matter of time before the cocklebur-encased tag adhered to the creature.

The team was up early on the 25th and discovered that, yet again, the string trap on which tag 7 was suspended had been compromised. The string was down, however, the tag remained; it had not adhered to whatever had disturbed the trap. The men reset the tag. They were now certain that an animal of substantial size was walking the path between the north and south cabins on a regular basis. Surely, the men posited, that ape cannot escape being tagged forever.

Two days later, the team awoke to the sound of something pillaging their food supplies on the front porch. Bass opened the cabin door to find a juvenile black bear pulling a plastic trash bag off of a post. Bass and the others made repeated attempts to frighten the bear off but they were unsuccessful. Finally, Haithcoat grabbed his shotgun and fired a round of buckshot over the bear's head. That did the trick, and the bruin hastily retreated back into the west woods.

By 7:50 p.m. on Friday, the 28th, Alton Higgins had arrived at camp. The team briefed him on the ongoing situation with tag 7. After unpacking, Wheatley and Higgins decided to visually inspect all of the string trap/tag sites. Upon arriving at the site of tag 7, they found the trap compromised and the tracking device missing. The pair double-timed it back to the cabin and began scanning. They immediately picked up the signal; the strong beeping tone emitted by the receiver indicated it was close. But by now, it was all but dark so the men decided to delay the search for tag 7 until first light.

The search began promptly at 8:45 the next morning. Oddly, the team could not acquire the signal as they had the night before. They double-checked the trap location; the magnet hung on a limp string but the tag was not there. Tag 7 was missing, but had gone silent. Either the tag had been destroyed by the animal to which it had adhered, or the animal carrying the tracking device had moved out of range of their scanning equipment. For the remainder of the summer, NAWAC teams attempted to re-establish contact with tag 7. These efforts are documented in the next chapter.

At 5:15 p.m., on Wednesday, September 2, Tony Schmidt and Higgins, now part of Lark Team, heard "two loud, impressive, primate-like howls, four to five seconds in duration" to the east. "They were among the loudest, most impressive vocalizations I have heard in the valley," said Higgins. Both men felt the howls had an "angry quality" about them. Just ten minutes later, Higgins, who was concealed between the south and west cabins, watched as a white-tailed deer "ran for its life down the creek bed" with a large canid of some kind in hot pursuit. Higgins believed it to be a wolf, even though wolves had long been thought extirpated from the region. It was large, much bigger than a coyote, had exceptionally long legs, a big head with oversized rounded ears, and carried its tail straight out while running.

On Friday, September 4, the team decided to stake out the area near the west cabin and beyond the property gate near the mud hole. The men stayed in place nearly four hours and heard multiple wood-knocks and anomalous crashes but saw nothing. At 5:00 p.m., Schmidt and Higgins began hiking back to the base cabin. Ed Harrison, who had arrived that morning, remained concealed in the brush south of the main road into camp.

Just three minutes after Harrison's teammates had entered the compound property, though he had no way of knowing this, Harrison observed two upright figures "moving briskly from west to east in a swale southwest of the RV." One figure was brown, the other was gray. He saw both swinging their arms as they walked but noticed their heads did not move or bob. "They walked with a fluid movement," Harrison said. The gray creature was "at least half a head taller" than its brown counterpart. Initially, Harrison believed he was observing his teammates but quickly realized that was not the case based on what he knew his fellow investigators were wearing and carrying.

Later, after recreating the event, the team determined that the two animals Harrison had observed were walking in a swale between the creek and a berm. The largest of the two subjects appeared to have been "close to seven feet in height." The distance between Harrison and the subjects was measured at 50 yards. They also determined that anyone walking down the main camp road—as Higgins and Schmidt had done on their way back to the compound—could not be seen from Harrison's vantage point. Whatever he had seen, it had definitively not been his two teammates.

The remainder of the summer operation featured several brief visuals, dozens of rock-throws, wood-knocks, and "weird vocalizations." The teams also spent a great deal of time attempting to locate the missing tag 7 and maintaining the other string traps. In addition, Kathy Strain, of Nightjar Team, had a startling visual of a large canid of some kind in the early afternoon of September 17. "It was larger than a dog," she said. "It had a bushy tail. It was reddish with a yellowish tail." This encounter was eerily reminiscent of Alton Higgins's sighting of a wolf-like animal a few weeks prior. It seemed there was room in the valley for more than one mystery animal.

Operation Resolute concluded on Friday, September 25 when the final NAWAC investigators of the summer locked the property gate and left the valley.

19
Tag 7

Early in 2015, NAWAC leadership began looking to expand its bag of tricks and find new methods which might help the group secure definitive evidence of the existence of the wood ape in Area X. We all realized that only a type specimen would completely satisfy mainstream scientists but we also suspected that other types of evidence—if compelling enough—could pique the interest of established scientists enough to get them involved in the search.

About this time, wildlife biologist and NAWAC investigator John Perry, of Maine, suggested we explore the use of radio tag and radio telemetry tracking devices as a means of collecting data on the target species. Perry had come to appreciate the capabilities and usefulness of this technology in tracking animal movements and wondered if it could be used to further improve our understanding of wood ape ecology and behavior. NAWAC Chairman Alton Higgins, who is a wildlife biologist himself, thought the technology had potential and could aid in the collection of significant ancillary data—*if* a means of deployment could be devised. Initially, the issue of deployment seemed to be too big to overcome, but Higgins realized that the solution to this issue might not require a radical new plan. Actually, the answer might lie in a research technique he had been using for years.

Higgins had begun deploying "string traps" in June of 2012 as a method for tracking ape movement. Based on the same concept as camp perimeter alarms, it involved deploying black sewing thread at a height of six to eight feet between two trees along suspected travel paths, one end tied securely to a tree, the other wrapped around a tree or limb in such a way that it would "unwind and stretch out in the direction of travel" if something were to walk through it. By positioning the thread so high, Higgins could eliminate the possibility of investigators or other more common forms of wildlife, such as deer or bears, coming into contact with the trap. The black thread was all but invisible to the naked eye, and on multiple occasions Higgins had found the threads compromised. He was convinced that the most likely creatures to have walked through the threads were wood apes.

So in March of 2015, Higgins and investigator Mark McClurkan decided to combine the two totally disparate methods of tracking animal movement—radio tags and string traps—into one novel tactic. The organization purchased two Advanced Telemetry Systems, Inc. (ATS) Model R410 scanning receivers, one three-element antenna, one five-element antenna, one omnidirectional Yagi antenna, and seven R1680 glue-on radio tag transmitters.[86] McClurkan was put in charge of figuring out how NAWAC personnel could use Higgins's string traps to get one of these radio tags attached to a wood ape.

How to get an ape to tag itself? As we previewed in the last chapter, the initial step involved gluing the small transmitters (designed for reptiles) into one-half of a spiny fruit from a cocklebur plant. The cocklebur is found across the North American continent but is especially prevalent across the south and has been the bane of humans and animals alike for as long as anyone can remember. Anyone who has found his/her legs covered in the spiny buggers after a hike or has tried to extricate them from the tail or mane of a horse or the fur of a long-haired dog can attest to this fact. The cocklebur seemed to be the perfect delivery system for the tags as they would almost certainly quickly become entangled in the hair of the target species.

McClurkan would then secure metal loops to the tag and its attached magnet; the tag would begin transmitting only when separated from the magnet. One piece of thread would run through the loop on the tag and be arranged in the way of a normal string trap. A second thread, shorter than the

Nano tag suspended across a game trail.

first, was tied to a tree or branch at one end and tied to the loop on the magnet on the other end. This set-up would ensure that the magnet would detach from the tag, which would then begin transmitting on a pre-set frequency, once it adhered to the moving animal. Each cocklebur would receive a coating of rat trap glue to help ensure adhesion to the subject and promote entanglement in the subject's hair. Rat trap glue was chosen for the job because it retains its extreme stickiness for months, even after exposure to the elements.

Once activated, a tag can transmit for 300-350 days before its battery expires. The signal can be picked up by the receiver and antennas at distances of up to three to five miles in flat, unobstructed terrain.[87] Since the range of these tags would be significantly less than that in the mountainous and thickly wooded Area X, investigators realized they would have to get very close to a tagged animal in order to pick up the radio signal. Everyone had high hopes that this new strategy would yield exciting and novel data about the habits of the wood ape.

During the summer operation of 2015, the tags were deployed in and around the cabin compound where wood apes had been seen over the previous years. On August 28, while NAWAC investigators in the valley were making their late afternoon check of the various string traps, they found that tag 7 was gone. As I reported in the previous chapter, the team quickly acquired the signal being transmitted by the missing tag, but because darkness was quickly setting in, the team decided to wait until morning before pursuing the tagged animal. When they did so the following day, their attempt to re-establish contact with tag 7 was unsuccessful.

Multiple attempts made over the next several weeks to regain contact with the missing transmitter were also unsuccessful despite investigators hiking up, over, and through hostile terrain for miles in all directions in an effort to re-acquire the signal. The efforts continued through the end of the summer to no avail. It seemed tag 7 was lost for good. Whether the tag had malfunctioned or been destroyed by the tagged animal, hopes dimmed quickly on ever being able to successfully track down the tag.

Just when the project seemed destined to fail, John Perry stepped in again. Knowing we had several pilots in the organization, he suggested attempting to acquire the signal from the air. Securing access to a suitable plane, getting that plane outfitted with the proper antenna, and coordinating pilot and ground team schedules while hoping for favorable weather conditions proved to be a logistical challenge, but the group was able to get multiple flights in the air beginning in December of 2015.

The air teams were, indeed, able to re-acquire the signal transmitted from tag 7 multiple times over the next several months and mark the locations via GPS. Each time a plane went up, a ground team was ready to deploy to the area of the signal in the hopes that the tagged animal could be tracked down via the use of one of the handheld antennas. The ground teams did manage to get close to the tagged animal on several occasions based on the strength of the signal acquired by the receiver; however, they were never able to lay eyes on their quarry. The rocky and vertical nature of the terrain made effective pursuit of the tagged animal on foot impossible.

The NAWAC tracked tag 7 from August of 2015 through June of 2016 using a method called discontinuous radio tracking. The technique involves locating an animal at discrete or random time intervals throughout the study period. The method, if a valid number of contacts are made, is useful in determining home range size.[88] In ten months, the NAWAC made contact with tag 7 a total of 25 times. While this is on the low end of acceptable sample sizes for home range assessments, once the difficult circumstances under which the animal was tagged and tracked were accounted for, 25 data points were considered adequate.[89] Location data were analyzed using Ranges9, software designed for the analysis of tracking and location information. Three methods were chosen to calculate home range, including the simplest, most traditional, and straightforward of range calculation methods. The Minimum Convex Polygon method (MCP) estimated a range of 18.75 square miles, the Adaptive Kernel Method estimated a range of 42.71 square miles, and the Ellipse Method estimated a range of 71.52 square miles.[90] NAWAC leadership wondered how these numbers stacked up against the known home-ranges of animals native to the Ouachita Mountain region.

Of the known wildlife in the area, the black bear is thought by many to be the likeliest candidate to have been the carrier of tag 7; however, it would have needed to be a larger-than-normal specimen to reach the transmitter, which was deployed 7.5 feet high. Even if a bear could reach the tag, it would likely have done so by pawing at the device. It is highly unlikely that the tag would have remained stuck to or near a bear's paw for a significant length of time. Besides, studies have shown that the range of black bears in the Ouachitas is much smaller than the range exhibited by the carrier of tag 7. Estimates range from 5.6 to 8.1 square miles for sow bears in the region (MCP and AKM methods).[91] Boar bears in the White River, Arkansas, area were found to range about 22 square miles (MCP method).[92] The range of the tagged animal tracked by the NAWAC appears to be much larger than that of a black bear. Perhaps the best argument against a black bear being the tag 7 culprit is the fact that the time during which the animal was tracked corresponds to the denning season for black bears in the Ouachitas. While bears do emerge for short periods of time during torpor, they do not range miles away from their den sites. The carrier of tag 7 moved too far, too often to have been a denning black bear.

Some have suggested a mountain lion as being the tag 7 carrier. The home range for the tagged animal does fall within the home range profile of the cougar, but the species is described as "uncommon" in greater Oklahoma by state officials, and there is no acknowledgement of residence at all in southeastern Oklahoma.[93] A "lack of residency" may be an invalid reason to discount the species outright, however, as NAWAC members have seen mountain lions in Area X. A better argument against the mountain lion as the culprit is its coat. The hair of a cougar is short and smooth and would likely not hold a tag for long. Mark McClurkan had tested the cocklebur traps on the coats of various species during the developmental phase of the project. He found "only momentary adhesion" took place on the coat of a deer. It is reasonable to expect similar results in regards to the likelihood of adhesion to the short-haired coat of a mountain lion. Finally, the idea that a cougar, a species famous for its furtiveness, boldly approached within 40-50 feet of a human encampment then leaped 7.5 feet into the air at the exact moment it passed under the string trap seems so unlikely as to not warrant serious consideration.

Skeptics of the wood ape hypothesis have suggested that a bird may have picked up tag 7. There are avian species in the valley that could conceivably have flown through the trap and are of sufficient size to have had the transmitter stick, owls, diurnal raptors, pileated woodpeckers, and wild turkeys among them. However, the extremely dense canopy in the location where the tag was deployed, as well as rough, thick understory and spacing of the trees between which the string trap was set make this improbable. Also, during the development phase of the project, McClurkan found that feathers did not adhere to the cocklebur traps, even when they were coated with rat trap glue. This fact, along with the small home-ranges and seasonal migration patterns of these avian species, would seem to eliminate some type of bird as the carrier of tag 7. Another flying animal, the bat, was also discounted. The study site is home to a large number of bats, but all either migrate out of Oklahoma or hibernate during the winter.[94] Whatever was carrying tag 7 was present and very active throughout the winter.

Canids and raccoons are often the last two species suggested as possible tag 7 carriers. Foxes have indeed entered the encampment over the years; coyotes have not. But neither of these species has ever been seriously considered as a suspect; not only are they unable to reach the 7.5-foot height of the trap but their home-ranges are significantly smaller than the carrier of tag 7. A raccoon could conceivably have climbed up one of the trees on which the trap was deployed and dislodged the set-up, but the large home range of the tag 7 carrier completely eliminates a raccoon as a serious suspect in this matter.

To summarize, over a period of several years, which includes thousands of hours in the field, animals fitting the description of wood apes, or Sasquatch, have been visually observed by NAWAC members. Some visuals took place in locations where string traps bearing small radio tag transmitters, modified to adhere to fur, were subsequently placed in the summer of 2015. One such tag was activated in late August of that year. Radio tracking over ten months suggested a large home range of up to 70 square miles. The fact that the tag remained adhered to the animal so long suggests the presence of thick, long hair. Additionally, the transmitter was deployed at 7.5 feet in height, which suggests that a very tall subject was tagged. The quick movement of the tagged subject through rugged terrain difficult for humans to navigate suggests an animal fully adapted to its environment. After much consideration, no type of known indigenous wildlife, domestic livestock, or feral species appears likely to have been the carrier of tag 7.

So if not a bear, mountain lion, bird, deer, coyote, fox, or raccoon, what walked through that high string trap in August of 2015? What did NAWAC investigators chase all over the Ouachitas that winter?

What?

20
Operation Fortitude

People often ask why, if these animals are real, has no one yet been able to prove their existence? There are many possible explanations for the lack of irrefutable proof, but I believe the main reason may be that humans are not capable of keeping drama, personal agendas, and/or their egos out of the equation long enough to achieve success. The early Bigfoot researchers, including John Green, Rene Dahinden, Peter Byrne, and Grover Krantz, among others, had their work affected, and at times crippled, by personal animosity, jealousy, and suspicion. Unfortunately, members of the NAWAC soon found out that they were not immune to these same issues.

Just as the organization seemed poised to achieve its greatest success, drama struck in the form of a falling out with the owner of the base cabin from which the organization had been working for the previous four summers. To this day, I remain amazed at how silly the disagreement was and cannot quite grasp how it got so out of hand so fast. The end result was that the NAWAC lost access to the cabin and the compound.

Group leadership decided to make the best of the situation by trying a completely different strategy. They selected a remote campsite on an elevated "saddle" between two rich hardwood-filled valleys that each had a healthy creek system running through them only four to five miles from the old cabin compound. The campsite was dubbed Blue 3 and was almost the complete opposite of the scenario the group had enjoyed at the old cabin. Blue 3 was on top of a ridge—approximately 1,930 feet in elevation—where the old compound was located on a valley floor. No longer would the apes have the high ground. Members were eager to see how this situation might affect the locals and if they would still feel comfortable approaching camp.

Since no cabin was available, the group purchased a high-quality Sibley 600 Ultimate canvas tent that was roughly 20 feet in diameter with a center ceiling height of 12 feet. This new "home" was christened the "Four Seasons" and, in many ways, was an improvement over the moldy, ramshackle cabin in which members had slept over the previous four summers. In addition, member Bill Coffman constructed a triangular-shaped frame out of PVC piping that would serve as the skeleton for a new overwatch tent. With a new base camp established, Operation Fortitude kicked off on May 14, 2016.

Apollo Team, consisting of Daryl Colyer, Jay Southard, Paul Bowman, and Ed Harrison, immediately went about the work of finishing the overwatch structure and putting the finishing touches on base camp. They also began scanning in the hopes of picking up a signal from the missing tag 7. On Wednesday, May 18, their efforts were rewarded when contact with the missing nano tag was re-established. The signal, one to three bars in strength, was picked up while the team was scanning a watershed valley to the south of camp. The tagged animal was near, and the men were thrilled that the transmitter was still functioning.

At 2:00 a.m. on Thursday the 19th, the men heard a large animal of some kind crashing through vegetation to the east, cross Blue Road (the moniker given to the ATV trail that led into the new campsite) from north to south, and head down the slope toward the valley floor and the creek located there. Five hours later, Harrison picked up multiple "jarring" signals ranging from three to four bars in strength while scanning for tag 7. The animal, whatever it was, was now in close proximity to camp. Intermittent "hits" were picked up over the course of the day and evening, and the team heard one "strange roar." But they saw nothing.

Despite a promising start, things slowed down and remained quiet until July 24 when Kraken Team members John Hairell and Phil Burrows encountered something odd while setting up an observation post away from base camp. Just before 4:00 p.m., the pair heard a "large tree crack and crash to the ground about 200 yards to the north." Twenty minutes later, the men heard a strange sound very close to them. "It was really odd...not like any bird I've heard," Burrows said. "It was two lower-pitched whistle notes about one second in duration with a one second pause in between them. It was puzzling." Later, the men heard what they described as "rock-clacking" that went on for several minutes. At 7:00 p.m., Leeper, who was alone back at base camp, heard "four distinct metallic strikes that were highly reminiscent of banging noises team members have heard back in the old compound for years." The team later combed through the woods in the area from whence the sounds had originated but found no metal of any kind. Leeper wondered, "Have the apes finally found us?"

Two days later, Leeper and Hairell were out for a hike after lunch when they heard "strange chattering sounds" in the forest. The men were amazed as they had traveled only 100 yards or so from the Four Seasons. "The sounds changed pitch and I am certain we were hearing two separate individuals," Leeper wrote. "One was deeper and more akin to ape sounds; the other was high-pitched and had a feminine quality." The men attempted to pursue the noise-makers, but the animals continued to give ground and move away. When the men would halt their pursuit, the chattering would begin anew. When the men tried to move in closer, the creatures would go silent and retreat. This cat-and-mouse game went on for two hours before the frustrated pair trudged back to camp.

When Hairell decided to hike the area again a few hours later, he was rewarded by the sound of movement in the brush just off a game trail near camp. He had hardly taken a step when he heard a "gruff and guttural huffing sound" quickly followed by the "crashing of a large animal through the brush." Hairell drew his sidearm and pursued the animal. After a few minutes, he came upon a thick wall of vegetation through which it seemed the creature had fled. The axiom, "Two is one, one is none," flashed across his mind, and Hairell decided not to pursue the creature into such a dense and restrictive environment alone. Chastened, he returned to camp.

Things once again went quiet at Blue 3 and stayed that way. Icarus Team, comprising Daryl Colyer, Travis Lawrence, Bill Coffman, and Jordan Horstman, arrived on site the second week of July and decided to try something different, sending a reconnaissance team on a long-range, multi-day excursion into one of the two valleys north of Blue 3. As they traveled, the men would attempt to elicit contact with the target species and locate tag 7. Over the next day or so, the away team heard odd vocalizations, a "strange, high-pitched growl," and wood knocks, but thermal scans showed nothing.

On the afternoon of the 12th, Colyer began blowing his "dying jack rabbit call" in an attempt to lure in an ape. Within minutes, Colyer caught sight of something big and black behind a tree approximately 40 yards to the west. It would turn out to be a sow black bear and her two cubs, a potentially dangerous situation.

The men quickly gathered their things and started hiking back toward their campsite, giving the bears a wide berth as they did so. Suddenly, the mother bear bolted from the brush and charged directly at the men. Colyer and Horstman raised their rifles to the high ready position and yelled loudly in an attempt to discourage the angry bruin. The bear stopped roughly ten yards from the pair and stared at the men. She seemed to have no fear of them at all. Just as it seemed she was about to turn and return to her cubs, the sow turned and began moving towards the men again. Colyer fired a single round above the bear. She began shaking her head wildly and fled back in the direction of her cubs. It would be the last the men saw of her.

Nothing else of note was documented that week.

NAWAC

4/09/09 7:35 AM

Black bear in similar pose to that of the sow bear spotted by Daryl Colyer and Jordan Horstman.

The remainder of the summer was largely uneventful. There was the occasional wood-knock, rock-clack, or odd smell from time to time, but activity at Blue 3 never came close to reaching the levels the group had experienced in the old compound. The group did, albeit briefly, re-establish contact with tag 7, which allowed them to add priceless points of contact to its collection of data and reach an acceptable number of "hits" from which an estimated home range could be extrapolated. This estimated home range all but eliminated bears, coyotes, cougars, or any other species known to inhabit the Ouachitas as the carrier of tag 7. And the group proved, even if only to itself, that members were not prone to misinterpreting sounds or misidentifying common animals. In other words, NAWAC personnel were not "seeing a wood ape behind every tree."

While the apes did seem to cross the saddle where Blue 3 had been locat-ed, they did not seem to tarry in the area for long and moved quickly across and down into one of the valleys on either side of our campsite. This seemed to confirm the hypothesis shared by most members that the apes lived in the valley(s) and, for whatever reason, did not spend much time in the higher elevations. It was clear that if the NAWAC was going to successfully achieve its goal of documenting the wood ape, the group needed to get back into the valley.

The question was: How?

21
Operation Dauntless

Fortune smiled upon the NAWAC when leadership managed to secure a multi-year lease on a remote piece of private property in the valley. The property, dubbed "Camp David," was only a mile away from the old cabin compound. The property was nestled up against the north-facing slope of the south mountain, and the main creek flowed through the valley nearby. The entire set-up was logistically reminiscent of the position of the north cabin in the old compound. The membership was excited and morale was sky high going into the summer of 2017.

Daryl Colyer, Bob Strain, Phil Burrows, and Tony Schmidt kicked off Operation Dauntless upon their arrival at Camp David on Saturday, May 20. That first night the team heard a clear "whoop-like" vocalization. When Schmidt produced a whoop of his own, immediately there was a response by what all agreed must have been the same individual. Schmidt produced another whoop in reply and the unseen animal answered again. This time the vocalization sounded much more like the whoop Schmidt had produced, as if the animal was attempting to mimic the investigator's call. This back-and-forth between Schmidt and the unseen visitor was repeated once more before the creature went silent.

The next night a black bear meandered into camp. It circled the Four Seasons several times before stopping in front of the tent entrance and rearing up on its hind legs. Realizing a potentially disastrous situation was at hand, Strain, who was in the overwatch structure, was ready to fire if it attempted to enter the tent where his teammates were sleeping. Fortunately, the wayward bruin thought better of it, dropped down to all fours, and walked out of camp.

At 4:00 a.m. on Tuesday, May 23, Strain, again while conducting overwatch, heard a strange vocalization. "It was a short run of garbled speech," he said, "a short pause of two to three seconds, then a shorter garbled 'Huh-weeahhuh-huh-huh...'" Strain fully believed he had heard an ape engaging in what the NAWAC has labeled "faux speech." Whatever was responsible for the sound had remained well hidden as Strain's thermal scans revealed nothing.

After a quiet day, the team blasted a mix of different primate vocalizations across the valley. The strategy paid immediate dividends when they heard what sounded very much like a "chimp hoot" from no more than 100 yards to the northwest. At 10:00 p.m., something hit the metal roof of the hooch with quite a bit of velocity. (The hooch was a metal carport structure open on three sides with a metal wall on the south side facing the mountain; it was designed to keep supplies and investigators dry during the many thunderstorms that roll through the region during the summer months.) The sound produced was much louder than the sounds made by the falling hickory nuts the team had heard periodically throughout the day. At 10:45, a "very primate-like whoop/scream" rang out from the mountain slope. The caller seemed close, no more than 75 yards away. Colyer answered with a call of his own and was rewarded when the unseen visitor sounded off again. This exchange was repeated four more times before the animal went quiet. "Those had to have been produced by either a human or an ape," said Burrows. "Nothing else can do that."

Due to how human-like the calls had sounded, the next morning the team looked for any sign of human presence on the slope. They found none. At one point, Colyer stopped about 100 yards up the slope and produced vocalizations similar to those heard the night before. His teammates below said his calls sounded as if they were coming from about the same distance as those heard the previous night. "Well then, there is no way on God's green earth that humans were up on this slope in the middle of the night attempting to hoax us," Colyer said. "The slope is slick and treacherous. There are boulders, loose rock, and greenbriar everywhere. There are hundreds of places to break an ankle, leg, or neck should some fool attempt to navigate up here in the middle of the night without using a light."

The remainder of the team's stay was not uneventful, though nothing as dramatic as the calls from the slope had occured. The team left the valley on Saturday, May 27.

Shortly after arriving on the morning of the 27th, Bravo Team was overcome by a "rancid garbage odor" wafting into camp. Initially, the team thought Alpha Team might not have hauled out their trash or had failed to empty the latrine. Neither turned out to be true as there was no garbage in camp and the latrine was found to be in perfect condition. Other than a couple of possible howls at 1:30 a.m. on Monday, and multiple whoops at 12:07 a.m. the next day, the week was relatively quiet.

Charlie Team, made up of Gene Bass, Jerry Hestand, Mark Porter, and Hans Helm, arrived on site on Saturday, June 3. The only possible ape activity documented by the team was a rock strike to the top of the hooch on Tuesday, June 6, and some eerie, raspy screams or howls on Friday the 9th. Most of the team bugged out the next morning. Only Helm remained at Camp David to await the arrival of Delta Team—NAWAC Chairman Alton Higgins and Jay Southard—on Saturday, June 10.

Two days later, in the wee hours of the morning, the men heard "multiple loud impacts on the hooch" along with rock displacement and other anomalous movement on the slope and woods near the latrine. Thermal scans proved fruitless, but later that morning the team discovered that string traps 14 and 15 had been compromised; both had been deployed by an earlier team near the latrine area where movement had been heard the previous night. Higgins felt sure that something had passed through them during the night. Unfortunately, neither of these two string traps had been outfitted with nano tags.

Later the men would find that string trap 5 west of camp had been snapped. The team quickly reset all compromised string traps but did not rearrange the location of the nano tags until Wednesday afternoon, June 14. Upon testing the range of the tags in the dense forest, they found the maximum distance at which the trackers could be detected was only about 200 meters. "Well, that's a bit discouraging," said Higgins.

Loud bangs and impact sounds were heard on Thursday the 15th. The next morning, the men discovered that string trap 5 had been snapped and trailed off seven feet in a westerly direction. "This particular string had been deployed at a height of six and a half feet," wrote Higgins. "Unfortunately, the radio tag did not adhere and remained in place. The string used here was very strong and we do not believe it possible for a bird to break it. After much discussion and testing of the string, such as pulling on the slender tree to which it was tied, and observing the same tree during strong gusts of wind, we were absolutely convinced that wind did not play a role in the breaking of the string."

The men re-deployed the string and tag, and at 5:00 that afternoon string trap 1 was found compromised. Like trap 5 earlier, the tag failed to stick to whatever had walked through it. While encouraged that something was disturbing the various string traps, the men were chagrined that none of the nano tags had successfully adhered to the animal responsible.

Echo Team, which arrived on site on Saturday, June 17, was plagued by severe storms their first few days in the valley. At 10:00 a.m. on the 19th, Travis Lawrence conducted a scan and found that tag 5 had been activated; it turns out the tag's magnet had become detached during the storm. While let down by the false alarm, the men took note of the fact that the string trap itself had remained intact through the fierce weather of the night before. Just after 10:00 that night, the team heard a "strange scream" from the northeast followed by a second scream. "I've never heard anything like that before," Lawrence remarked to Helmer.

When the team reviewed the audio recordings from the previous night, they were elated to find the odd screams had been successfully captured. But upon close scrutiny of the vocalizations, elation turned to concern. "The screams might be more accurately described as roars," Helmer wrote. "They seem much more menacing when played back than when we heard them in person. Disturbing." Helmer dubbed the vocalizations "T-Rex roars." (The "T-Rex Roar" audio file can be heard on the book's webpage at anomalistbooks.com.)

Wednesday the 21st and Thursday the 22nd featured "some sort of chatter" east of camp, a "hudd-wah" sound from the same vicinity, and other "pops, wood-knocks, and a whistle." Just before lunch on Friday, the 23rd, the men were standing under the hooch talking when they heard a "very loud bang" on the roof followed by some bouncing sounds. Upon investigation, Lawrence found "a large rock" in the spot where the bouncing object had come to rest on the roof. It was larger than most rocks previously collected by members, measuring 4 x 3 x 1 inches. It had struck the metal roof with enough force that some of the structure's green paint had adhered to it.

On Tuesday, July 4, Golf Team recorded a visual at 1:46 a.m. Leeper, who walked to the tree line just west of the hooch in order to relieve himself, caught sight of "green eye reflection" 20 yards to the north. The creature, whatever it was, bolted away to the northwest. The next morning, the team conducted a recreation of Leeper's experience from the night before and determined that the eyes had been about six and one-half feet off the ground. "It was quite a sobering realization," remarked Leeper.

Hotel Team—consisting of Jay Southard, Ron McCollum, Travis Lawrence, Ed Harrison, and biologist Angelo Capparella—arrived at Camp David on Saturday, July 8. Their primary goal for the week was to deploy two long-term autonomous recording units (ARUs) in the area around camp.

At 9:00 p.m. on Monday the 10th, the team tried a remote call-blasting technique that Capparella had successfully used in the Siskiyou Mountains of northern California. Capparella blasted three unnamed howls, each of which lasted an incredible 17 seconds. Two minutes later, they heard four clear return howls. The team tried additional playbacks without success.

Between 1:30 a.m. and 4:40 a.m. on Tuesday, July 11, Capparella, while sleeping in his own small tent away from the Four Seasons, heard something strike the side wall of his tent, a "low, quiet humming," multiple wood-knocks, a strange "thud," and movement in the vegetation around camp. Then he heard a "low growl" to the northwest "that sounded like the warning growl a dog would make when a person reaches for his/her food bowl." The activity ceased when the first rays of sunlight appeared.

Hotel Team gave way to India Team on Saturday, July 15. Alton Higgins, Bud Mellicker, and Jay Southard spent the first two days of their stay clearing a new area and repositioning the Four Seasons tent, which had flooded multiple times over the previous weeks. The team heard multiple impacts on the hooch roof over the next couple of days.

On Thursday, July 20, Higgins and Mellicker, while hiking to the west of camp found a dilapidated, old tree stand. Beneath it, Higgins spotted tracks that he was certain had been made by a wood ape. "There were eight to ten tracks total, one of which still possessed some toe impressions, leading away from the base of the hunting stand to the south," Higgins wrote. "The tracks measured approximately fourteen inches in length and were about one-half inch to one inch deep in the soil." After inspecting the trackway, Mellicker wrote that he was "quite impressed."

Tod Pinkerton, of Juliet Team, arrived at Camp David on Saturday, July 22. Pinkerton would be alone on site for the first few days of the week. His only visitor was a young black bear that wandered into camp on Sunday. Pinkerton successfully frightened the bruin away.

The most interesting, and unnerving, event of the week took place at 3:30 a.m. on Tuesday, July 25. Pinkerton awoke "to the sounds of faint footsteps and breathing" outside the west wall of the Four Seasons. "Something just wasn't right about the situation," he said. "I grabbed my .357 revolver and headlamp and quietly rolled over onto my back."

The footsteps ceased at that point, but the "breathing" sounds continued.

"The breaths occurred every five to eight seconds with each 'breath' lasting two to three seconds," Pinkerton said. Thinking he would not have the time to get to the tent flap, unzip it, exit the Four Seasons, and spot the intruder before it fled, he instead grabbed his keys and remotely unlocked his truck which was parked to the east of the tent and facing north. This action caused the headlights to temporarily engage and light up the east side of the tent. Pinkerton hoped the "breather" would be startled and make a false move; or maybe the creature would be silhouetted against the side of the Four Seasons, allowing him to identify the culprit.

Unfortunately, the maneuver, which he repeated several times, did not work. The "breather" was evidently not bothered in the least by the light. Pinkerton momentarily considered exiting the tent in an attempt to get a look at his nocturnal visitor, but ultimately decided against it. He was alone and did not know for sure what was lurking outside and what its intentions might be. Eventually, he would doze off only to be jolted awake at 6:30 a.m. by a loud bang somewhere outside of camp.

Kilo Team, made up of Daryl Colyer, Matt Pruitt, and Angelo Landrum, relieved Pinkerton on Sunday, July 29. Shortly after arriving, the men produced multiple wood-knocks, whoops, and broadcasted the sounds of a distressed kid goat in an attempt to "get the party started." They received no immediate response; but at 8:58 that night, the team heard what sounded like the cries of a frightened goat. The men stared at each other with disbelief; the sounds were amazingly similar to what they had call-blasted earlier. "Incredible," muttered Pruitt.

On Thursday, August 19, Landrum and Higgins were examining the roof of the hooch when they heard the familiar sound of a large tree cracking. The men watched, as the top 20 feet of a tree on the mountain slope snapped off and crashed to the ground. Upon investigation, they found that the tree had long been dead, which might explain the fall, but it did not explain the wildly violent to-and-fro swaying and shaking they had witnessed on a windless slope.

Camp David had been vacant for four days when investigator Dusty Haithcoat arrived on Thursday, August 24, as the first member of Oscar Team. He would spend the first two days alone as teammates Scott Wheatley and Tony Schmidt would not arrive until Saturday. Mere minutes after arriving at camp, Haithcoat heard an odd cadence of "clacks" coming from the northwest. "I'm sure it was the sound of two rocks being struck together in a rhythm of one impact every second," he wrote. This rhythmic clacking went on for a solid ten minutes. A few moments after the northwesterly "clacker" went silent, similar sounds reached Haithcoat's ears from the southwest, albeit from a greater distance. Like those from the northwest, the southwest clacking lasted approximately ten minutes before ceasing. *Party on*, the lone investigator thought to himself.

After a quiet night, Haithcoat first heard what sounded like "the loud exhalation of a very large animal between the Four Seasons and the hooch," then later the sound of what he was "certain were human voices from the northwest just behind me and beyond the hooch." Investigation yielded nothing, but later that afternoon while hunting along Camp David Road, Heithcoat heard what sounded like the garbled speech of a human male," Haithcoat said. "It had to have been within 50 yards." Again, there was nothing there. Then at 1:30 a.m. on Saturday, the 26th he heard a strange "monotone whistle." The animal was close, he guessed, within 30 yards. The whistle was repeated seven more times. "It was just…odd," said Haithcoat. Around mid-morning, Wheatley and Schmidt arrived.

The "whistler" returned on the morning of August 29. The entire team heard a whistle that "slid down in tone from high to low and was much louder than the birds singing at the time." Wheatley produced his own whistle in an attempt to mimic what they had just heard. The unseen visitor answered. This back-and-forth went on two more times before the whistling stopped. Minutes later, they heard an explosive sound as loud as a gunshot; a rock had blasted the roof of the metal structure. The team would experience several more explosive rock-impacts over the next several days before exiting the valley on Saturday, September 2.

Romeo Team was the final team of the season. Alton Higgins, Daryl Colyer, and Brian Brown arrived at Camp David on Wednesday, September 13. The usual shenanigans followed: the sound of a loud impact on the roof; the sighting of two illuminated green spots at the base of the slope initially thought to be bioluminescent insects but possibly the eyeshine of an ape; the sound of a nearby "growler;" the *thump, thump, thump* sound and *feeling* of something walking calmly away from camp; and what "sounded like a female human in the creek making a high-pitched wailing."

Less than an hour after those first wailing calls were heard, the creature vocalized again, this time from farther east. By now, the team had nicknamed the maker of the haunting wails "the banshee." The high-pitched call chilled the blood of even the experienced Brown and Colyer. It was no more than 100 yards away. In an effort to draw the animal even closer, Colyer answered with a wail of his own. The banshee answered once and then went silent.

On the night of Sunday, September 17, after hearing movement on the mountain off and on for hours, the team caught a glimpse of green eyeshine up on the slope southwest of the hooch. The men were startled as they were in a pitch-dark camp. No torches were lit, no headlamps or flashlights were on, and there was no fire in the fire pit. The only ambient light was that provided by the stars above. Colyer quickly glassed the slope with a thermal optic and spotted a large, white-hot heat signature. Brown confirmed the signature with his own thermal unit. It appeared to Colyer that he was seeing an animal peeking from behind a tree. The animal rocked back and forth, moving from its "peeking" position to one of almost total concealment behind the tree. The men noted "secondary signatures" on the opposite side of the tree that were lower than the main heat signature. The men agreed that they were likely seeing a hand and fingers, as if the animal was peering out on one side of the tree while holding onto the trunk with the hand on the opposite side. After observing the animal for several minutes, Brown moved toward the slope. The animal immediately "dropped" out of sight and was not seen again.

After a long day on Monday, the 18th, the team had retired to the Four Seasons, when they heard the sound of padded footfalls racing through camp in an east to west direction. As the visitor passed the hooch, it loudly slapped or struck the back wall. The men could not believe the audacity of the camp intruder. It had seemingly waited for the men to retire and then immediately sprinted through camp as if demonstrating that it could breach the perimeter of the NAWAC encampment at will. Brown and Colyer burst out of the Four Seasons in an attempt to catch sight of the "camp hoodlum" but they were too slow. Whatever it had been, it was long gone. As the pair stood in the dark scratching their heads, a sharp wood-knock rang out from the northwest.

The men were again awakened by the sound of something or someone in the camp at 5:20 in the morning of the 20th, but like before they were too late to catch a glimpse of the "runner." Since he was up, Colyer decided to walk to the edge of camp to relieve himself. As he did so, he called out "yoo-hoo" loudly. To the shock of both men, an unseen animal to the north replied "yoo-hoo," though it was less drawn out and higher-pitched than Colyer's. "This place is freaking nuts," mumbled a stunned Colyer.

The men broke camp and left the valley at 7:35 a.m. bringing Operation Dauntless to a close.

22

Operation Intrepid

Operation Intrepid kicked off the summer of 2018 when Daryl Colyer, Ken Helmer, Matt Pruitt, and Joel Thomas arrived at Camp David on May 19.

Early in the morning two days later, the team was shaken from sleep by Pruitt's car alarm going off. When Helmer sat up quickly and looked out his tent's screen door, he saw that the area around Pruitt's vehicle and the overwatch tent was illuminated, as the headlights had come on. As Pruitt scrambled for his keys to deactivate the alarm, Helmer saw an animal fleeing to the southeast.

The creature had been "lit up" by the headlights and was running hard for the slope of the mountain. Helmer was certain he had seen the top portion of a wood ape. "It was a big brown mass with some orange in coloration, high off the ground, perhaps eight feet or so." The remainder of the week would prove quiet, and Alpha Team exited the valley on Saturday, May 26.

Things remained quiet until Charlie Team, comprised of Angelo Landrum, Jay Southard, Mark Porter, and Caleb Huckins*, arrived on site. On Wednesday, June 6, Huckins was checking the string traps and nano tags when he spotted something moving through the brush. "It was about 30 yards away. It was a very large, upright animal that was reddish-brown in color," he said. "It was the approximate color of an Irish Setter, had shaggy hair, and was walking rapidly away from me toward the mountain slope." Though the sighting was brief, only about two seconds, Huckins noted that the "bright green background allowed for a perfect complementary color contrast." Huckins estimated that the animal was approximately eight feet tall and three feet wide. Huckins quickly hoofed it back to Camp David in order to alert his teammates.

Delta Team arrived at Camp David on Saturday, June 9 and constructed a small cabin, which would be dubbed the "Harrison House." Members were glad to learn that they would once again have access to the extra security and shelter provided by four solid walls and a roof.

Echo Team, consisting of Daryl Colyer, Brian Brown, Travis Lawrence, and Bob and Kathy Strain, arrived on Saturday, June 16. The next day, Alton Higgins arrived with filmmakers Seth Breedlove and Adam Duggan, who would be shooting footage for an upcoming Small Town Monsters production.

At 3:10 p.m., while concealed beneath a large hickory tree, Colyer spotted movement in the dense vegetation 90-100 yards away. The movement proved to be the swaying of several young saplings. At first, Colyer assumed they were being moved about by wind but soon noticed no other vegetation in the immediate area was swaying. He decided to glass the area with his rifle scope. When he did, Colyer saw "a flash of something white or gray" behind the brush. After watching the anomalous light-colored animal for several minutes, the investigator observed the upper portion of a "conical head with slitted eyes" raise up above the foliage and peer down in his direction. After a few seconds, the animal moved noisily away.

At 11:27 that night filmmaker Breedlove saw "two lights" in the trees up on the mountain slope just southeast of the new cabin. Colyer and Brown saw the "lights" as well. It was clear to all that what they were seeing was the greenish-colored eyeshine of an animal. Breedlove and Duggan again saw a set of "eyes" just to the west of their tent at 1:00 a.m.

The next afternoon, around 4:25, Brown, Breedlove, and Duggan heard a "perfect wood-knock to the northeast" while out checking string traps. Less than an hour later, Bob Strain heard "talking" from the east, followed minutes later by a "loud yell". The activity would culminate in the wee hours of Tuesday morning when the hooch was pounded by some sort of projectile, followed by a "gibbon-like whoop." Breedlove and Duggan, impressed with what they had seen and heard, exited the valley the next evening.

During what Colyer and Brown called a "staggered hike" to the east of Camp David on the afternoon of the 21st, Colyer decided to bang two rocks together. Within minutes, the sound of rocks being banged together erupted from the forest. The "banger" was close. Colyer banged his rocks together a second time. Again, the men received a clear and close response. Wide-eyed, the duo attempted to creep closer to the unseen noise-maker but never did lay eyes on the party responsible and returned to camp.

Meanwhile, Bob Strain had found a new nut-crushing station complete with a hammer stone and crushed nuts. "Hammer stone present," wrote Kathy Strain the next day after visiting the site. "Station has nuts on it and meat under the hammer stone. Station measures 32.25 inches long x 16.5

NAWAC

Nut-crushing station with hammer stone.

inches wide x 9 inches high. The hammer stone measures 6 inches x 4 inches x 1.5 inches."

The members of Foxtrot Team rolled into Camp David shortly after the exit of Echo Team. Phil Burrows, Caleb Huckins, Gene Bass, and Ed Harrison were tasked with erecting an omnidirectional antenna that would scan for deployed nano tags without the need to use one of the unwieldy handheld antennas. Nothing truly significant occurred until their seventh day, when just after midnight on Saturday, June 30, Burrows and Harrison heard the unmistakable sound of a substantial tree snapping and falling somewhere to their northeast. The men, headlamps on, turned in time to see a tree crashing to the earth. They then heard a large animal tearing through thick vegetation. The men sprinted toward the downed tree but were never able to see the animal.

On Sunday, July 15, I joined teammates Shannon Graham and Shannon Mason at Camp David. They had arrived a day earlier and had heard multiple "clear wood-knocks" around camp. The next morning, we stumbled across several possible footprints in a muddy spot on the road leading into camp. The impressions were not pristine but certainly appeared to have been made by something bipedal. I photographed the possible prints, but, as is often the case, the pictures did not do the impressions justice.

Minutes later Graham momentarily spotted a "small, shaggy, red-dish-brown animal" through a window in the vegetation. While a cinnamon-colored black bear could not be completely ruled out, Graham felt strongly that the "shaggy hair" did not fit the bear hypothesis.

That evening, Mason entered camp's solar shower and shouted "Whoo!" as the surprisingly cold water hit her body. Graham and I, sitting back at the dark fire circle, got a hardy laugh out of Mason's predicament. Our laughter turned to amazement minutes later when we all heard a very loud and very clear "whoooo" vocalization from the slope of the mountain just south of the hooch. I was stunned. It was easily the best vocalization I had ever heard in my many years of pursuing these animals. It was close, maybe as little as 30-40 yards away. When Mason answered our unseen visitor with another "whoo" of her own, the animal answered her, but changed pitch to better mimic the investigator's call. This exchange took place one more time before the creature went silent. The three of us looked at each other, slack-jawed. "Incredible," I muttered.

After a quiet Tuesday, our visitor returned at 9:16 that night. A "whoooo" rang out from the slope of the mountain. The vocalization was virtually identical to Monday's call and seemed to come from the same spot. As before, Mason called back. The animal replied once and then went quiet. Fortunately, we had a TASCAM audio recorder running and successfully recorded the vocalization. (The "Intrepid India Whoop" audio file can be heard on the book's webpage at anomalistbooks.com.)

Thursday morning, the three of us took a short hike to the west of camp but saw nothing of interest. The weather had turned incredibly hot—at one point the camp thermometer read 105° F—and the short hike wore us out. Mason began exhibiting the early symptoms of heat exhaustion and felt quite ill the next day. Though the army veteran was willing to tough it out, we decided to pull out one day early as the valley was a long way from help should Mason take a turn for the worse.

The next team to experience significant activity was Mike Team. The team, made up of Hans Helm, Caleb Huckins, and Jay Southard, located "ape-like prints" on Monday, August 13, in the same muddy area where Graham, Mason, and I had spotted suspected tracks a few weeks before. The tracks measured "16 inches long, 8.5 inches across the toes, 4.5 inches at the heel, and impressed 2-3 inches in the substrate. There were three tracks total."

On Thursday, the 16th, Southard and Helm went on a hike to the west while Huckins stayed at Camp David. At 1:56 p.m., Huckins heard a series of 20 loud bangs from one of the "banging stations" (pieces of corrugated metal strung between trees) that had been deployed by an earlier team. We knew, based on experiences in the old compound, that the apes often made loud noises by banging on scrap metal and wanted to provide the locals with the same opportunity near Camp David. The strategy appeared to be paying dividends as Huckins listened to the piece of metal to the west being pounded repeatedly. Roughly 20 minutes after the banging ceased, he heard a menacing "huff/growl/grunt." He wrote: "It was too loud to have been a smaller animal or a bird." Moments later, he caught a brief glimpse of a reddish-brown, hair-covered animal through the foliage to the northwest. Huckins watched in amazement as the animal silently slipped away.

Camp David had been vacant for two weeks when Romeo Team, comprising Daryl Colyer, Brian Brown, Matt Pruitt, Rick Major*, and Alton Higgins, arrived in the valley on September 15. The next morning at 5:04 the team reported hearing "faux speech" to the north; it sounded like "rah, rah, rah." While teammates investigated, Colyer and Major prepped for a three day, long-range, reconnaissance patrol that would take them to previously unexplored drainages to the west of camp. Higgins and Pruitt dropped off Colyer and Major at the predetermined insertion point at 8:00 the next morning. The away team members were now on their own.

That night, beginning at 10:20, the pair reported hearing knocking, tapping, popping, various odd vocalizations, rock displacement, and the sound of running footfalls off and on all night long. But whatever was out there managed to stay just out of sight and avoid detection. "We are being probed," Colyer wrote. "We went out to confront them and insulted their mothers. They backed off only to return. CLOSE wood-knock as I write."

At 2:45 a.m. on the morning of the 18th, Colyer and Major observed two large, green reflective eyes in the wood line. "The eyes blinked as the animal watched us," Colyer wrote. The creature backed off only after the pair of investigators hurled an expletive-laden rant its way. A bit later, after the men had managed to doze off, Colyer was awakened by a subtle noise just beyond the thin wall of their two-man tent. He then saw a "large shadow of something as it quickly passed by the north side of the tent." The shadow was possible only because he and Major had strung up low-intensity light-sticks outside before turning in for the night. Having had enough and in desperate need of sleep, Colyer got up and hung a bright lantern outside the tent; all activity ceased "as if a switch had been turned off."

A fully reunited team would go on to report hearing the usual odd sounds over their final few days in the valley. They heard mouth-pops, wood knocks, crashing trees, "long, moaning-type howls," *hoo* vocalizations, a two-toned whistle, and footfalls—and had a fleeting glimpse of a black animal darting across a dry creek bed.

The entire team exited the valley at 9:38 a.m. on Thursday, September 20, bringing Operation Intrepid to a close.

23

Operation Variance

The leadership of the NAWAC was encouraged by the level of activity experienced by its members during Operation Intrepid, but at the same time, frustration regarding the inability to "finish the deal" had taken root. One member said, "It is like hunting a shadow. How can these apes continue to make their presence known so often and still continue to elude us?" He went on to add, "I'm beginning to wonder if it is ever going to happen for us."

"Yes," I replied to the discouraged investigator. "The apes are incredibly elusive, but they have made mistakes, mistakes that could have easily ended in our acquiring the specimen we seek. They have been very lucky on multiple occasions. Sooner or later, that luck will run out. Remember, they have to be lucky *every* time; we just have to be lucky once."

We needed to vary our tactics and try something different. We decided to split field operations for 2019 into three different legs with scheduled breaks in-between when no team would be in the valley. Would this change lead the apes to relax their vigilance and prompt a slip-up on their part? We hoped so.

Operation Variance kicked off in 2019, but the spring leg proved to be quiet, maybe because the lack of cover (the hardwoods had not yet leafed up) forced the locals to keep their distance. The languor seemed to have carried over into the second leg of Variance after Delta Team experienced little in the way of ape-related activity the first week of June. But this period of inactivity would prove to be the quiet before the storm.

Echo Team arrived at Camp David on the afternoon of June 8. Alton Higgins, Phil Burrows, and Hans Helm made up the team. Two days later the team had several what might be called "sound encounters": a "deep-throated howl," a "sliding noise" like something had slid down a tree, a wood knock, and a "metallic impact sound."

On Wednesday, the 12th, the investigators again hiked out west of camp to set up in three different parts of the area where they believed the "emphatic wood-knock" had originated two days before. Despite the heat, Higgins and Helm donned ghillie suits for the exercise. Burrows did not have one so he wore a large brown poncho. At 7:15 p.m., Higgins spotted an upright figure walking in a southeasterly direction only 40 yards from his position.

The figure was brown, and Higgins initially figured he was seeing the poncho-wearing Burrows as he changed positions. Higgins quickly lost sight of the figure in the thick brush. The longer Higgins pondered what he had seen, the more it bothered him. Something just seemed "off" somehow. Higgins quickly gathered up his stool and rifle and walked toward the trail in an effort to catch up with "Burrows."

Meanwhile, Burrows, who was hunkered down just off the main trail, caught sight of the brown, upright figure. "It was standing still and facing south," said Burrows. "Initially, I thought I was seeing Alton." A few minutes later, Higgins walked up on Burrows and immediately saw that his teammate was no longer wearing the brown poncho. In fact, Burrows was not wearing anything brown. After checking on Helm, who had stayed far to the west, Higgins and Burrows were convinced they had seen a brown wood ape.

By Sunday, June 16, the members of Foxtrot Team were on site in the valley. Present were Daryl Colyer, Tom Smith*, Jim Ross*, Hans Helm, and Jay Southard. Things were quiet for the first two days, so on Tuesday, the 18th, Helm and Smith decided they would spend the night near the old tree stand where Higgins had located a trackway in 2017. That night, they were rewarded with two consecutive "mysterious moaning" vocalizations from the northeast. "It sounded like an old man moaning loudly out in the woods," said Smith. "It was creepy." The rest of the night would be quiet.

The next morning, the two investigators were making their way back to Camp David when they flushed a "very large animal" that had been lurking in the thick brush just off the trail. According to Smith, "The animal swiftly bolted through the woods from east to west. It crashed through the brush displacing many rocks as it apparently sought to avoid being seen. Whatever it was, it was a VERY LARGE animal." In his field notes, Helm wondered, "Could it have been an ape sentry keeping tabs on Colyer and Ross that was unaware Smith and I were inbound from the west?"

One of the more exciting events of the summer would unfold later that night. The team was running a dark camp; there was no campfire, lit tiki torches, or any other white light in use. The men were sitting around the cold fire pit when at 9:53 p.m., Ross and Southard heard brush movement and rocks sliding to the east. Ross, using a handheld monocular, immediately spotted "a large, amorphous heat signature." Colyer, who was peering through a thermal scope mounted on the overwatch rifle, confirmed the signature.

Colyer stood up in order to get a better view of the figure. "It was close, maybe 40-45 yards away," he said. "It was huge, skulking, and appeared to be creeping backwards." Colyer and Ross came to the same conclusion: they were seeing a "head and shoulders," potentially a wood ape whose thermal image was broken up a bit by vegetation.

As the figure continued its slow and steady retreat away from the team, Colyer saw what he was certain was an arm protrude upwards and grab or move a tree branch as the subject continued moving backward. Adrenaline rushing through his body, Colyer took his weapon off safe. His teammates' eyes widened when they heard the tell-tale click, but no shot came.

"It looks human-like," said Colyer. "I will not fire unless I get a much clearer shot and can definitively tell it is not a human being." Colyer was experiencing real inner turmoil. He had been in this situation before. His mind told him that he was observing a wood ape and that no human would be skulking around in a location as remote as Camp David in the middle of the night. His gut, however, told him he needed to be 100% sure he was not glassing a person as the figure had definite human-like characteristics. Seconds passed and the figure slowly melted back into the forest east of camp. Colyer lowered his rifle and looked at his teammates. "I just couldn't be sure," he said. They agreed he had made the right decision by not firing.

The next morning, the team recreated the scene from the night before and concluded that the figure had been close to seven feet in height. Colyer sighed, knowing now that he had observed an ape and an opportunity for collection had been lost. "Still," he said, "it is better if 100 apes get away than it is if we make a single mistake." Foxtrot Team would exit the valley on the 22nd.

Golf Team members Alton Higgins and Jerry Hestand joined holdover Jay Southard in the valley at 6:10 that evening. Two days later, at 11:00 p.m., while in the brush east of Camp David, Southard whispered to Higgins that he had acquired a strong heat signature back toward the cabin, where Jerry Hestand was resting up for overwatch later that night. Higgins swung around and quickly picked up the signature as well. The bright object was located near the northwest corner of the cabin, and whatever it was seemed to be using a porch post and a rain collection barrel as cover. Higgins kept the reticle on the head of the subject but never took his weapon off safe or placed his finger on the trigger. "We had left Jerry in that cabin," he said. "I *knew* what I was looking at was not Jerry; he would never behave in that way knowing Jay and I were out there scanning with the overwatch weapon. Still, no shot could be taken as Jerry *was* in that cabin. All I could do was observe as the animal slowly withdrew from sight."

NAWAC

A recreation of the Camp David "Porch Peeker" witnessed by Alton Higgins and Jay Southard.

Hotel Team, made up of Travis Lawrence, Alton Higgins, and Phil Burrows, had a few strange experiences during their stay in the valley, though they were subtler than the thermal visuals experienced by the two previous teams. At 4:35 p.m. on Sunday, June 30, upon returning to camp after hunting to the west for several hours, the men found a bright pink hair clip on the ground. "There's no way we would have missed noticing it if it had been there before we left camp," said Lawrence. Someone or something had visited camp in their absence.

The following Wednesday night, July 3, Lawrence and Burrows heard an "apey whoo whoo" vocalization that was quickly followed by the sound of a tree cracking. A minute later, the men heard a loud "ooo ooo ooo" vocalization that echoed across the valley. The men described the sound as "clearly a pant-hoot-type vocalization."

I, along with Tony Schmidt, Dusty Haithcoat, and Ron McCollum made up Lima Team. We arrived on site on Saturday, July 27. It would be one of the strangest weeks of my entire life.

Schmidt and I were resting in the cabin Sunday evening in preparation for overwatch when a loud bang from up on the slope south of the cabin was heard by the entire team. We were all in agreement that the sound had almost certainly come from the southernmost banging station on top of the ridge. It was as if the apes were announcing their presence. It would be only a prelude to the much more intense activity to come.

Late on Sunday night, while McCollum was in the cabin and Haithcoat was in his hammock, Schmidt and I were sitting near the dark fire pit when we were startled by strange sounds from just west of the cabin near the hooch. The sounds were quite unlike anything I have ever heard in the woods before. It sounded like someone beating the ground repeatedly with a large stick. You could hear the stick, or whatever it was, cutting the air just before striking the earth. This went on for nearly a full minute. Schmidt and I stayed in place and scanned the area west of the cabin and around the hooch but could not see who or what was responsible.

A few minutes later, the sounds started again. The noisemaker had changed locations and now seemed even closer. I would have sworn that whatever was beating the ground was *under* the hooch and mere yards from Schmidt and me. The impacts were louder this time, as if the striker was swinging its "club" even harder than before. *Whap! Whap! Whap!* Despite how close the sounds seemed, we still could not see the noisemaker through our thermals. The impacts went on for about 25 seconds and seemed to be increasing in intensity. Finally, convinced there was an ape under the hooch and that my thermal device could not possibly be working correctly, I clicked on my headlamp. There was nothing there, and the impact sounds stopped.

Schmidt and I were flabbergasted. How was it possible that we were not seeing the maker of these sounds? It sounded like it was *right there* and uncomfortably close. More frustrated and amazed than shaken at this point, we decided to try to lure the "ground banger" in close enough to be seen. We moved on to the porch of the cabin, hiding behind the black plastic of the overwatch panels. Within minutes the impact sounds began anew but they had greatly intensified. I find it difficult to express just how loud and powerful the ground-strikes were. Over and over again, the impacts were repeated. Finally, something struck the hooch with terrific force. Schmidt and I were stunned at how loud this impact was and how the hooch reverberated for several seconds afterward.

We immediately burst out from behind the black plastic panels and off the porch, scanning with our thermals and then with white lights. We saw nothing. Whatever it had been, it was now gone. "How could it have been so close and remain concealed?" I asked. Schmidt had no answer for that. The rest of the night was uneventful. This incident remains the single strangest and most intense experience I have had during my nearly two decades of research.

Schmidt, McCollum, and I left the valley on the morning of Tuesday, the 29th, in order to procure the hardware necessary to lock the new door to the cabin. I will not pretend that the intimidating display from Sunday night did not factor into our decision to get the locking mechanism functional as quickly as possible. That afternoon, Haithcoat, who had elected to stay in camp, heard a series of bangs he described as sounding like "someone repeatedly using a stick to beat on a five-gallon bucket." He heard six different sets of bangs in a three- to four-minute period. Eight minutes later, the banging started again. This time he noted three sets of rhythmic strikes. And again 12 minutes later, Haithcoat documented yet another round of impacts, the last of which increased in speed and reached a crescendo of nearly continuous banging.

On Wednesday, the 31st, we decided to try an experiment that we hoped would lead to a collection opportunity. Just after lunch, McCollum and I hiked out of camp toward our teammates who had positioned themselves about 100 yards apart in the dense thicket off a well-worn game trial to the east about an hour before. The hope was that an ape would key in and follow us, inadvertently exposing itself to Haithchoat or Schmidt in the process.

We had donned bright orange vests so our hidden field partners would know without a doubt whether they were seeing us or some other bipedal creature. As we walked down the trail and arrived at the dry creek bed, we found Haithcoat standing out in the open with a perplexed look on his face. "I think I just saw an ape," he said. His account, paraphrased from our personal conversation, is as follows:

"I looked to the north toward the trail and saw a five- to six-foot-tall figure covered in black hair standing in the creek bed. It was sort of leaning into the west bank wall of the creek and appeared to be peeking up over the edge, as if observing something on the level of the forest floor. I couldn't be absolutely sure of what it was I was seeing. It was lean, upright, and had black hair. The hair on the head was sort of spikey-looking, like Ernie from *Sesame Street*. I could not make out a face or any other distinguishing features. I wanted to glass the animal, but was in a bad position to do so. I took my eyes off the animal momentarily to shift around and bring my rifle up to a position where

I could examine it through my scope. When I looked up again, it was gone. I did not hear it move as it left. It had just vanished. I left my hiding spot to walk to where I had seen it and that is when you guys (McCollum and I) came walking down the trail."

We discussed the possibility that Haithcoat had seen a black bear, but he was sure—since he saw the animal in profile—that he would have seen a snout had it been a bruin. Besides, he said, "This thing did not act like a bear. It was stealthy and absolutely silent. I really think, whatever it was, that it was aware you two were coming down the trail and was attempting to get a look at you." It seemed our plan had indeed worked, but instead of something following McCollum and I and walking past our hidden teammates *after* us, the animal was already in the area and was *ahead* of us on the trail. This event cemented in our minds the fact that these animals do make mistakes. If the animal Haithcoat observed was an ape, and I strongly believe that it was, it screwed up. It had no idea Haithcoat was there and was vulnerable.

The next notable activity took place during the stay of Mike Team, which arrived in the valley on September 14. Five days later, investigator Jody Blaylock, who was sleeping outside the cabin in a tent, was awakened at 5:07 a.m. by a loud wood-knock from the south. A few minutes later, a tree crashed to the ground in the northeast, waking Alton Higgins who was sleeping on a cot on the cabin porch. Less than an hour later, they both heard a deep "Hrahh!" vocalization from the east. It seemed at least one ape was in the area.

At 9:35 that morning, one of the more alarming events of the summer occurred when investigator Tod Pinkerton was struck in the head by a large hickory nut. Momentarily stunned, Pinkerton sank to the ground. Chad Dorris, who was nearby at the time, was sure he had observed the nut fly in horizontally from the woods to the west. If so, the nut had to have been *thrown* as nuts falling from a tree do not move horizontally. Pinkerton was not seriously injured, but the incident remains one of only a handful of times when a NAWAC investigator was intentionally targeted by a thrown object.

November Team, consisting of Daryl Colyer, Brian Brown, Rick Major*, and David Haring, arrived on site on September 21. That night the team was greeted by "a high-pitched 'aah' that lasted about two seconds," a substantial projectile that blasted the hooch just ten minutes later, and finally a series of vocalizations "like a sustained garbled yelling," Brown said.

On Monday the 23rd, Major, having taken a shower using the rainwater collection system located behind the cabin was dressing when he heard a commotion up on the slope of the mountain to the south. He immediately spotted a large, reddish-brown animal in a draw on the slope that appeared to be observing him from a crouched position. Immediately, the creature burst from its hiding spot and fled to the southwest and up the slope at a high rate of speed. "Bloody hell!" Major cried as he watched the animal flee. The other men came running, and although they all heard the animal crashing through the brush as it made its escape, they were unable to lay eyes on it.

The next day, the team staged a recreation of the shower incident. Brown stood in for the ape and squatted in the draw where the ape had been hunkered down. Major quickly realized that the animal he had seen dwarfed Brown. Any lingering doubts as to what he had seen melted away. "I believe I saw a wood ape," he said.

The team would hear the usual assortment of odd sounds over the next two days, including wood-knocks, mouth pops, and rock-strikes. Perhaps the oddest was a "yahoo" vocalization that Brown heard three times at 1:00 a.m. on the 26th while camping at what the team called "Monkey Hollow," an area of interest to the west of Camp David.

Jeremy Lindsey

Investigators exploring the jungle-like area near "Monkey Hollow."

The final event of significance to take place during Operation Variance occurred at 9:46 on the morning of Thursday, September 26. The group had loaded into Brown's truck to go pick up some food supplies, when Brown, rounding a bend in the road out of the valley, caught sight of an upright, charcoal gray figure standing on the west side of the road. Brown slammed on his brakes and watched the animal move off to the west and into the wood line. The men hastily exited the vehicle to pursue the creature. But they quickly realized that attempting to follow the animal through the thick woods and across the steep terrain was folly. Brown was certain what he had seen was a wood ape, possibly Old Gray himself.

The team would exit the valley the next day. Operation Variance had come to an end.

24

A Pause and a Pivot

The work of the NAWAC was greatly affected by the Covid-19 pandemic in 2020 and 2021. Due to concerns about the health and safety of members, most of the field activities scheduled for these two years were canceled. In addition, the idea that Covid-19 could be passed from humans to wood apes did give us pause; the last thing anyone wanted was to expose the resident apes to a deadly virus. The Sasquatch is likely our closest living relative in the animal kingdom, possibly sharing up to 99% of our DNA (as a reference, chimp and human DNA is 98.8% identical).95 The NAWAC exists to conserve this magnificent species, and the thought of possibly harming a regional population of these rare animals by inadvertently introducing this new virus was anathema to us.

That said, an extremely abbreviated summer field exercise was planned, and Operation Trinity kicked off on July 11, 2020. The handful of weeks during which NAWAC members occupied the valley were mostly quiet and uneventful. The one exception was a visual by investigator Lawrence Burns, who caught a brief glimpse of a dark-colored, bipedal figure moving in a northerly direction just west of the latrine area of Camp David on July 22.

NAWAC leadership decided to give trail cameras another try in 2020. The project, dubbed "Hadrian's Wall," was designed to photograph any and all wildlife that walked through the valley. Dozens of cameras were posted in a sort of picket line across the valley floor in a camera-watching-camera fashion. The hope was that improved camera technology would enable us to capture the compelling images/footage we seek. The project has yet to bear fruit, but we remain undeterred and will be adding cameras to the array soon.

During the summer of 2021, ad hoc teams made up of members at a low risk of infection periodically visited Area X. While some ape-related activity was documented, nothing of a truly dramatic nature took place.

The organization has now purchased some new high-end thermal devices. Not only do these new thermals have improved resolution, they also have the ability to take video. This is something our old thermal devices and scopes could not do. If the apes remain active near Camp David, new and compelling thermal images should be captured during a forthcoming operation.

The years of the Covid-19 pandemic have provided the NAWAC with a period of reflection. During these two years, the Board of Directors decided to reevaluate its organizational goals and priorities. The stimulus to do so has been the realization that the NAWAC efforts to date have been, in the eyes of many, a failure. *Why was the NAWAC unable to obtain a specimen in over ten years of trying?* Most seem to think that since we "see these things all the time" it should have been easy. The reality is that we do not see apes "all the time." Not by a long shot.

Between 2011 and 2021, NAWAC members reported 51 "hard" sightings of wood apes. (By hard sightings, I mean visuals that could not have been anything else; the "soft" sightings of large animals *suspected* of being apes have been removed.) On the surface, this seems like a very large number. Once the data is crunched, however, a different truth is revealed. During those ten years, the NAWAC deployed dozens of teams, ranging in size from one to six members, to the field. All told, this equates to roughly 778 days of observation. When the number of days in the field is divided by the number of visuals, it equates to 15.2 days, between sightings. Add the fact that nearly all visuals were a mere two to three seconds in length and it becomes clear as to why a specimen proved so difficult to obtain. Rarely were these short windows of time enough to clearly identify the animal *and* raise a weapon.

I would add that NAWAC hunters have also demonstrated a vast degree of self-discipline. On multiple occasions, members conducting overwatch have observed compelling heat signatures through thermal scopes mounted on rifles but held their fire. Almost without exception, the animals observed were partially obscured by thick foliage, making a positive identification impossible. Strict protocols, brief visuals, thick vegetation, the incredible elusiveness of our quarry, and an abundance of caution have led to only five occasions when shots were fired over the last ten years: the Echo Incident of 2011; the bluff charging incident experienced by Mark McClurkan in 2012; the "Iron Man" incident of 2012; the Kilo incident of 2013; and Tony Schmidt's "rock-peeker" incident of 2015. Five incidents in a decade. That's it.

I feel strongly that this fact should dissuade the narrative that NAWAC members have been running around the woods "playing commando and shooting at anything that moved." It just is not true. I am every bit as proud of the shots not fired during our time in the valley as I am of the few that were taken. The bottom line is that sightings are rare and short in duration which makes getting a shot at an ape a ridiculously difficult task.

Which leads to the most common question asked of the NAWAC: *Why, after all these years, have you been unable to capture any convincing photos or video?* It is a fair and valid question. The answer is very simple: we have not been trying. I am not being glib in any way; the truth is that the NAWAC has long believed that no photograph or video will ever be good enough to constitute irrefutable proof that the wood ape exists. That being the case, most members simply have not been focused on obtaining this type of evidence.

But now the Board of Directors has come to realize that opportunities to collect valuable evidence have gone by the wayside due to our single-minded pursuit of a holotype. Tracks have gone uncast, unmeasured, and unphotographed; the search for secondary evidence like hair and/or scat has been mostly an afterthought; and no attempts to collect environmental DNA have been made. While none of these types of corroborative evidence would have been enough to definitively prove that the wood ape is real, perhaps—had we continued to prioritize their collection as we had in the early days—the organization would now be able to make a more compelling argument for the existence of the Sasquatch in Area X.

The realization that potentially valuable evidence was going uncollected was deemed unacceptable, and when paired alongside the now obvious difficulties of hunting these creatures, it was decided that a shift in priorities was in order. The Board of Directors realized that had many of those members engaged in hunting been carrying cameras instead of rifles, convincing images would likely have been captured many times over. After all, a hunter must exercise extreme caution and take the time to clearly identify his/her target before sending a round. A photographer has no such limitations and can simply snap away without hesitation.

With this in mind, group leadership mandated that the acquisition of photographs and video be the number one priority of the NAWAC moving forward. Make no mistake, I and the vast majority of the membership still feel we will need an actual physical specimen to positively prove the existence of the wood ape to science. After a decade of trying, however, we have concluded that we simply do not have the manpower or resources to make collection a reality. That being the case, we hope to build a portfolio of photographs and video so convincing that mainstream science will have no choice but to sit up, take notice, and join us in our efforts to solve this mystery.

25
What Have We Learned?

Although the NAWAC's efforts over the last decade have not resulted in concrete proof of the existence of the Sasquatch, I do not believe these ten years have been wasted. I believe we have accomplished much. The organization has compiled more than a decade's worth of field notes documenting the behaviors of these amazing creatures. The publication of the *Ouachita Project Monograph* was a huge accomplishment. Similarly, the use of radio telemetry tags to track what the group strongly believes to have been an ape, and the subsequent publication of the "Tag 7 Paper"—officially titled "Tracking a Self-Tagged Unidentified Species in the Ouachita Highlands"—solidified the fact that the NAWAC is dedicated to scientific methodology.

Moreover, we have learned much about the behavior of these animals. But just what we have learned is not an easy question to answer. I will attempt to address the most asked-about behaviors as succinctly as possible.

When I first got involved in Sasquatch research in 2005, it seemed nearly everyone believed these animals to be solitary in nature. NAWAC work has revealed—at least regarding the apes that populate Area X—that is not the case. Where there is one animal there is likely to be at least one other nearby. Members have documented wood-knocks, pops, whistles, and clicks from multiple locations within seconds of each other. There have also been multiple sightings of apes walking in pairs. Investigators have also been subjected to distraction tactics when closing in on one of these animals. Often, when in pursuit of one creature, a second ape will vocalize, strike a tree, thrash vegetation, or throw objects in an attempt to draw the attention of the pursuers away from the other ape. Membership now operates under the assumption that if one ape reveals its presence, there is likely another one nearby.

I believe we now have a pretty good handle on how these animals react under duress as well. The longer investigators stay in ape territory, the more overt will be the attempts to drive the invaders out. Wood-knocks lead to rock-throws which escalate to whoops or howls if the interlopers refuse to leave. If those tactics fail to achieve the desired results, intimidation displays like growling, slapping cabins, and bluff charging tend to take place. I think the evidence strongly suggests the "aggression" often reported by witnesses is nothing more than an attempt to get unwelcome visitors to leave the area. The escalation of behaviors does seem to have a cap, however, as the apes do not seem to have much more in their bag of tricks once these options have been exhausted. NAWAC data seems to indicate the temperament of these animals is akin to that of orangutans in that they are not prone to engage in violent behavior with humans. As evidence of this, I would point to the handful of times when investigators have shot at these animals. Even in the midst of these extreme circumstances, the apes in question simply retreated.

There have also been times when I am firmly convinced the Area X apes were toying with us for nothing more than the sheer fun of it. I am not saying that wood apes are harmless by any means. I believe they deserve the same healthy respect one would give to a black bear or cougar when encountered, as any animal can become dangerous when cornered, afraid, or protecting young. But they do seem hard-wired to avoid humans, not attack them. It is possible that there are some bad apples in the Sasquatch bunch that might do a person harm, but I feel they are few and far between.

NAWAC researchers have come to believe that there is a breeding population of wood apes in the Ouachitas, in and around Area X in particular. Sightings in the region go back to at least the 1840s. These animals simply could not continue to be seen into the present day unless they were successfully breeding. Witnesses and NAWAC investigators have reported seeing small ape-like animals in the valley and surrounding areas. There is the gentleman who related seeing a small, upright animal in a tree 60 yards from his tree stand while hunting in the valley (as related to Alton Higgins in 2001); the small black ape Daryl Colyer saw sprint across the trail in 2012; the four small, chimp-like animals Travis Lawrence observed dropping out of a tree and retreating from him in 2014; and the small animal with an unusual gait spotted by Higgins in 2014. These observations all serve as anecdotal evidence that juvenile apes inhabit the valley at least some of the time.

The NAWAC has also learned that wood apes do, on occasion, produce "faux speech." Like many of my fellow investigators, I was once dubious about much of the behaviors that others had attributed to Sasquatches. Chief among these alleged behaviors was wood-knocking (which I now absolutely believe to be ape-related) and a type of vocalization that other researchers called "Samurai chatter." After listening to some very well-known recordings of these types of vocalizations, I felt they were the most ridiculous sounding things I had ever heard. I have had to swallow some crow on that front as I, and many other NAWAC members, have heard very similar sounding "chatter" over the years. I remain unconvinced that this faux speech represents an actual language, but I do believe it is communicative in nature and one of the many unusual vocalizations these animals are able to produce.

Then there is the wood apes' reputed ability to mimic certain sounds and to whistle. Based on observations made over the last decade, I believe wood apes are able to whistle and mimic a surprising number of sounds. The observations documented by NAWAC personnel jibe nicely with Native American lore regarding the whistling ability of the Sasquatch. The idea that wood apes can whistle is not as unlikely as it may at first seem. Humans can whistle, as can orangutans.[96] The wood ape would be just one more primate with this ability.

Of greater interest, at least to me, is the ability of these animals to mimic a wide variety of sounds. NAWAC members have heard an amazing assortment of sounds that defy traditional explanation. Among them would be the sound of something attempting to replicate the popping and buckling of the base cabin's corrugated tin roof as Alton Higgins walked on it in 2011; the unseen creature that seemed to be recreating the sound of a saw cutting a tree as reported by Daryl Colyer, Brian Brown, and Brad McAndrews in 2011; and whatever exchanged "whoops" with Tony Schmidt in 2017 and Shannon Graham in 2018. Other reported auditory anomalies include the sounds of cars starting, freezer lids slamming, and goat-like bleating. I think too much credit is sometimes given to these animals when it comes to mimicry—most often an owl is just an owl—but, having said that, the evidence collected by the NAWAC seems to corroborate the ability of these apes to replicate a wide range of strange sounds.

Of equal importance might be what the NAWAC has *not* observed during our decade in the valley. No member has ever reported any mind-speak or telepathic communication with one of the locals. Neither has anyone been "zapped," observed a portal open or close, or watched as a wood ape suddenly "cloaked" or morphed into a shimmering, translucent being as seen in the movie *Predator*. In fact, the members of the NAWAC have encountered nothing that could even remotely be considered paranormal or make us reconsider our position that the wood ape is a flesh and blood animal. I know there are those out there who have suggested that the very reason we have failed to obtain a specimen or capture compelling photographs is due to the esoteric nature of the Sasquatch. I tend to think that the reason we have yet to obtain concrete proof has more to do with rugged, inhospitable environments and a nimble, furtive, and incredibly intelligent adversary.

But the NAWAC has not stopped trying. Stay tuned.

26

A Few Last Remarks

Some people will no doubt dismiss the accounts in this book as one huge fabrication: a hoax. It's true that the history of Sasquatch research is filled with shady characters and outright charlatans. It's one of the reasons the topic has been shunned by mainstream science. But the truth is, no matter what I say, I cannot absolutely prove to anyone that the incidents described here actually occurred. But not one of the more than 75 current and past members of NAWAC has ever claimed any one of our accounts was a hoax. It goes without saying that some members have exited the organization with hard feelings. But while a few former members have made unflattering remarks about the organization, *no one* has ever claimed that anything the organization has said, written, or recorded regarding our observations, evidence collected, or research results is a fabrication.

Then there are people who feel that NAWAC members are telling the truth about what they have experienced in Area X but believe *we* are the victims of hoaxers. I suppose I can understand why someone who has never been to the valley might harbor such suspicions. But the notion that hoaxers have spent three months a year, for more than a decade, hiding in the rugged and mountainous Ouachita Mountains in order to throw rocks at cabins, produce an occasional vocalization, fake trackways, and traipse about in an ape suit to torment, tease, and fool a group of heavily-armed NAWAC members (who, until recently, have been bent on collecting a specimen) is not only ludicrous but potentially suicidal. In addition, many of the more intense ape-related incidents to have taken place over the years have occurred at night when the presence of one of the apes was detected only with the help of a thermal device. The idea that a hoaxer could, or would, navigate a steep, rocky, and vine-covered mountain slope in the dark of night without using any kind of light simply cannot be entertained.

While ape-related activity in the valley has continued over the years, it is certainly not as intense as it was during those first four summer operations. There could be multiple reasons for this ebb in activity. The apes inhabiting the valley are unquestionably incredibly intelligent. While we have learned much about them, they have learned much about us, too. The locals have learned many of our tricks and ruses and are not as easily fooled or goaded into revealing their presence as they once were. For lack of a better term, the apes of Area X have been "educated." I would be willing to bet that no wood apes anywhere else in the world have been hunted or pursued to the extent of those in that valley.

A second possibility for the slowing of activity is that the apes may have simply grown accustomed to our presence and are no longer as intensely curious about us as they were those first few summers. It has been posited that the apes that have so often tormented team members with rock-throws, growls, the slamming of freezer lids, and the occasional slap of a cabin wall were young. Perhaps they were the equivalent of human teenagers who feel invincible and so often chase the adrenaline rush that accompanies risky behavior. Maybe these young apes have matured and no longer feel the need to be so bold as often.

Our change in location might also be part of the equation. The cabins in the old compound have been there for generations; they are part of the environment. Camp David is new and as such, might be a place deemed risky or scary. Perhaps the apes will ratchet up their shenanigans once the shine has worn off the new location.

In any case, my fellow NAWAC members and I feel a true sense of urgency to capture the evidence we seek and move the wood ape from the realm of myth to that of a scientifically-described and documented animal. Only then will governmental protections for the species and its habitat be afforded. Only then can the future of this magnificent North American primate be assured. Again, development is not slowing, and more habitat is lost every day. This species must have large tracts of forestland to survive. How long does the wood ape have until there is simply no longer enough suitable habitat to sustain a breeding population? How long does the species have until that tipping point is reached and the extinction spiral begins?

We need to succeed in documenting this species before it's too late. Only then will I be able to rest in the knowledge that Area X and its inhabitants are safe. Then, finally, we can all be assured that Area X will forever remain the Valley of the Apes.

Acknowledgements

One does not complete a project of this magnitude alone. This book simply could not have been finished without the help and cooperation of a great number of people. I would like to express my deepest gratitude to each of you who contributed to this effort. Please know that I am immensely thankful for your friendship, patience, prodding, and input.

This list of special people includes the past and current members of the NAWAC. You are truly a remarkable group of individuals. Joining your ranks was one of the best decisions of my life. Though some of you have moved on to other things, you all hold a special place in my heart. I will forever consider each and every one of you a brother or sister.

Daryl Colyer, your efforts to solve this mystery have been nothing short of Herculean. Without your meticulous compiling of more than a decade's worth of field notes, authoring of scholarly papers, and blunt insistence on adhering to scientific protocols and methodologies while serving as Field Operations Coordinator, none of this would have been possible. I hope I can help you fulfill the promise you made to your fellow USAF airmen all those years ago and solve this mystery once and for all.

Alton Higgins, your even-keeled leadership has been nothing short of remarkable. You have steered the NAWAC through good times and bad with equal aplomb. It has been my honor to know and serve with you all these years. If I can be half the chairman you were, I will count my tenure as a great success.

Finally, I want to thank my wife, Holly. Simply put, you are the greatest blessing in my life and I am thankful you choose to find my quirky interests endearing. I love you.

Sound File Appendix

The NAWAC sound files available to the public. Each of these files was recorded during one of the Area X field studies chronicled in this book. Some of the subtler sounds are best heard through headphones or earbuds; however, most of the audio can be heard clearly using computer or phone speakers only. These files were originally made available when the *Ouachita Project monograph* was published in March of 2015.

Sound files can be accessed at:
https://www.woodape.org/index.php/opmonograph/#audio

Audio clip 1: A wood-knock just before dawn just outside the base cabin.

Audio clip 2: A rock hits the cabin followed by the sound of a very loud wood-knock.

Audio clip 3: Huffs outside of the cabin followed by the sound of a rock striking then rolling off the roof.

Audio clip 4: Huffs outside the cabin followed by a rock striking the structure.

Audio clip 5: Ape-like huffs followed by another rock-strike.

Audio clip 6: Various sounds: a whistle, shuffling, huffing, and a rock-strike.

Audio clip 7: Huffs followed by the sound of a rock clipping vegetation before slamming the cabin and bouncing to the ground.

Audio clip 8: The sound of a single huff followed by a rock zipping through foliage and striking the cabin.

Audio clip 9: A rock is heard clipping foliage before landing with a thud short of the cabin.

<u>Audio clip 10</u>: While the team was talking in front of the cabin, a rock flies through the trees and strikes the cabin.

<u>Audio clip 11</u>: A rock hits the loose corrugated metal on one of the sheds near the base cabin and bounces away.

<u>Audio clip 12</u>: The "rain of rocks" event. While the team was in the cabin trying to rest, the cabin was continuously bombarded by rocks big and small. The barrage lasted for several minutes. This is an abridged version.

<u>Audio clip 13</u>: The sound referred to in this book as a "mouth pop" or "tongue click." This is the sound Mark McClurkan mimicked before being charged by a wood ape while stationed at OP1.

<u>Audio clip 14</u>: Field audio documenting what has been variously described as "mumbling," "gibberish," "chatter," and "faux speech."

References and Citations

Chapter 1

1 Sanderson, Ivan T. "The Strange Story of America's Abominable Snow-man." *True Magazine*, Dec. 1959.

2 Ibid.

3 Coleman, Loren. *Bigfoot: The True Story of Apes in America.* Paraview Pocket Books, 2003, p. 67.

4 Dobie, J.F. *Tales of Old-Time Texas: J. Frank Dobie*. Little, Brown and Co., 1956, pp. 14-33.

5 Arment, Chad. *Historical Bigfoot.* Arment Biological Press, 2006, p. 314.

6 Ibid., 316.

7 Alvarez, Elizabeth Cruce & Plocheck, Robert. *Texas Almanac-The Source for All Things Texan Since 1857: 2008-2009. The Dallas Morning News.*

8 Colyer, Daryl, and Higgins, Alton. "Wood Ape Sightings: Correlations to Annual Rainfall Totals, Waterways, Human Population Densities and Black Bear Habitat Zones." *North American Wood Ape Conservancy.* https://www.woodape.org/index.php/ecological-patterns/. Dec. 12, 2007.

9 Treat, Wesley, et al. *Weird Texas: Your Travel Guide to Texas Local Legends and Best Kept Secrets.* Sterling, 2009, p. 92.

10 Ibid., 96.

11 "2014 Louisiana Forestry Facts." *Louisiana Forestry Association*, http://economic-impact-of-ag.com/LA/2014LA_Louisiana_Forestry_Facts-20rylte.pdf

12 Bord, Janet, and Colin Bord. *Bigfoot Casebook Updated: Sightings and Encounters from 1818 to 2004.* Pine Winds Press, 2006, p. 219.

13 "The Legend of the *Loup Garou*." *Night of the Loup Garou*, http://mississippiwerewolf.com/Night_of_the_Loup_Garou_Movie_Website/The_Legend.html.

14 Nadel, Darcie. "The Rougarou: Louisiana's Cajun Werewolf." *Exemplore*, 25 Jun 2020, https://exemplore.com/cryptids/The-Rougarou-Louisianas-Cajun-Werewolf.

15 "Honey Island Swamp Legends." *PearlRiver Eco-Tours*, http://www.pearl-riverecotours.com/legends.

16 Ibid.

17 Abbott, B. Nick, and Richard A. Marston. "Ozark Plateau." *Oklahoma Historical Society*, https://www.okhistory.org/publications/enc/entry.php?entry=OZ002.

18 Cole, Shayne R., and Richard A. Marston. "Ouachita Mountains." *Oklahoma Historical Society*, https://www.okhistory.org/publications/enc/entry.php?entry=OU001.

19 "Oklahoma's Forests." *Oklahoma Forestry Services.* https://forestry.ok.gov/oklahoma-forests.

20 Bord, Janet, and Colin Bord. *Bigfoot Casebook Updated: Sightings and Encounters from 1818 to 2004.* Pine Winds Press, 2006, p. 218.

21 Ibid., 37.

22 Ibid., 227.

23 Ibid., 47.

24 Ibid., 233.

25 "The State of Arkansas." *Netstate*, 28 July 2017, https://www.netstate.com/states/intro/ar_intro.htm.

26 "Arkansas's Forest Facts." *Arkansas Department of Agriculture.* https://www.agriculture.arkansas.gov/wp-content/uploads/2020/05/2017_Forest_Facts_of_Arkansas.pdf

27 Blackburn, Lyle. *The Beast of Boggy Creek: The True Story of the Fouke Monster.* Anomalist Books, 2013, pp. 58-59.

28 Mangiacopra, Gary S., and Dwight G. Smith. "Wild Men and Mountain Gorillas: A Historical Retrospective of 19th Century Sasquatchery Encounters as Recorded in North American Newspapers." *CRYPTO Homology Special Number II*, 2002, pp. 40–47.

29 Arment, Chad. *Historical Bigfoot.* Arment Biological Press,, 2006, p. 49.

30 Ibid.

31 Ibid., 50.

32 Ibid., 50-51.

33 Bord, Janet, and Colin Bord. *Bigfoot Casebook Updated: Sightings and Encounters from 1818 to 2004*. Pine Winds Press, 2006, pp. 9-10.

34 Reel, Monte. *Between Man and Beast: An Unlikely Explorer, the Evolution Debates, and the African Adventure That Took the Victorian World by Storm*. Doubleday, 2013.

Chapter 3

35 "Sam Houston National Forest." *National Forests and Grasslands in Texas*, USDA, https://www.fs.usda.gov/detail/texas/about-forest/districts/?cid=fswdev3_008443.

36 Colyer, Daryl, and Alton Higgins. "Operation Primate Web II." Texas Bigfoot Research Conservancy. Lorena, Texas, Jan. 2005.

37 Ibid.

38 "Howls Recorded in Cuyahoga Valley National Park (2015)." *The Bigfoot Field Researchers Organization*, http://www.bfro.net/avevid/SOUNDS/cvnp_ohio_howls.asp.

39 Colyer, Daryl, and Alton Higgins. "Operation Primate Web II." Texas Bigfoot Research Conservancy. Lorena, Texas, Jan. 2005.

40 Ibid.

41 Griffin, Andrew. "Local Bigfoot Buff Resumes Search for Elusive Beast." *The Town Talk*, 24 Mar 2005, p. D2.

42 Ibid.

43 Colyer, Daryl, and Alton Higgins. "Operation Primate Web II." Texas Bigfoot Research Conservancy. Lorena, Texas, Jan. 2005.

44 Griffin, Andrew. "Local Bigfoot Buff Resumes Search for Elusive Beast." *The Town Talk*, 24 Mar. 2005, p. D2.

Chapter 4

45 "The Big Thicket." *National Parks Service*, U.S. Department of the Interior, https://www.nps.gov/bith/index.htm.

46 "The Biological Crossroads of North America." *National Parks Service*, U.S. Department of the Interior, https://www.nps.gov/bith/learn/nature/index.htm.

47 Ibid.

48 Ibid.

49 Colyer, Daryl, and Alton Higgins. "Operation Thicket Probe." Texas Bigfoot Research Conservancy, Lorena, Texas, Sep. 2005.

50 Hammond, Keri, et al. "Weird Travels." *Weird Travels*, season 1, episode 14, Travel Channel, 6 Jan 2006.

Chapter 5

51 Colyer, Daryl. "Operation Thicket Probe II." Texas Bigfoot Research Conservancy. Lorena, Texas, Jan. 2006.

52 Higgins, Alton. "Operation Thicket Probe I and II Commentary." Texas Bigfoot Research Conservancy. Oklahoma City, Oklahoma, May 2006.

Chapter 6

53 Fields, Johnathan. "Robot Watches Out for Woodpecker." *BBC News*, 18 Feb 2007, http://news.bbc.co.uk/2/hi/science/nature/6372911.stm.

54 Ibid.

55 Pradeep, K. "Man of Nature." *The Hindu*, 1 Oct 2006, https://www.thehindu.com/todays-paper/tp-features/tp-sundaymagazine/man-of-nature/article3232638.ece.

56 Casstevens, David. "Sasquatch Watch." *Fort Worth Star-Telegram*, 31 Oct 2005.

57 "Managing attractants." *Get Bear Smart Society.* https://www.bearsmart.com/live/managing-attractants/

58 "Photos: A Timeline of SETX Hurricanes." *Beaumont Enterprise*, 31 May 2017, https://www.beaumontenterprise.com/news/article/Photos-A-timeline-of-SE-Texas-hurricanes-11185848.php#photo-8078749.

Chapter 7

59 Bader, Christopher D., et al. *Paranormal America: Ghost Encounters, UFO Sightings, Bigfoot Hunts, and Other Curiosities in Religion and Culture.* New York University Press, 2010, p. 124.

60 Colyer, Daryl, et al. "NAWAC Investigators Hear a Number of Close Knocks and Discover Fresh Tracks the Next Day." *North American Wood Ape Conservancy, https://reports.woodape.org/data/?action=details&-case=01080009#CaseNum.*

61 Bader, Christopher D., et al. *Paranormal America: Ghost Encounters, UFO Sightings, Bigfoot Hunts, and Other Curiosities in Religion and Culture.* New York University Press, 2010, p. 125.

62 Ibid.

63 "MonsterQuest - Swamp Stalker." *MonsterQuest*, season 3, episode 3, History, 18 Feb 2009.

64 Ibid.

65 Ibid.

Chapter 8

66 "Mission Statement." *North American Wood Ape Conservancy*, https://www.woodape.org/index.php/mission-statement/.

67 Butler, Rhett A. "Photos: The Top New Species Discoveries in 2012." *Mongabay*, 26 Dec 2012, https://news.mongabay.com/2012/12/photos-the-top-new-species-discoveries-in-2012/.

68 Ibid.

69 Erickson, Jim. "Collecting Biological Specimens Essential to Science and Conservation." *University of Michigan News*, 22 May 2014, https://news.umich.edu/collecting-biological-specimens-essential-to-science-and-conservation/.

70 "Is Current Specimen Collection Justified for Scientific Purposes?" *ResearchGate*, 8 Aug 2012, https://www.researchgate.net/post/Is-current-specimen-collection-justified-for-scientific-purposes.

71 Ibid.

72 Ibid.

73 Van der Pluijm, Ben. "Deforestation." *Resilience.Earth,* University of Michigan, https://resilience.earth.lsa.umich.edu/units/deforestation/index.html.

74 Ibid.

75 Fuller, Errol. *Lost Animals: Extinction and the Photographic Record.* Princeton University Press, 2014, p. 17.

76 Green, John. *Sasquatch: The Apes Among Us.* Hancock House, 2006, p. 384.

77 Geggel, Laura. "Top 10 Things That Make Humans Special." *LiveScience,* 02 Feb 2022, https://www.livescience.com/15689-evolution-human-special-species.html.

78 Green, John. *Sasquatch: The Apes Among Us.* Hancock House, 2006, p. 381-382.

79 Ibid., 382.

Chapter 9

80 Roberts, Roger, and Alton Higgins. "Report #4532." *Bigfoot Field Researchers Organization*, 18 June 2002, http://www.bfro.net/GDB/show_report.asp?id=4532.

81 Higgins, Alton, and Brett Elliott. "Fall 2001-Southern Oklahoma-Ouachita Mts." *Bigfoot Field Researchers Organization*, Oct. 2001, http://www.bfro.net/avevid/ouachita/opreport.asp.

82 Ibid.

83 Ibid.

Chapter 10

84 Colyer, Daryl, et al. *"Ouachita Project Monograph." North American Wood Ape Conservancy*, 2015, https://www.woodape.org/index.php/opmonograph/.

85 Ibid., 16-17.

Chapter 19

86 Bowman, Paul, et al. "Tracking a Self-Tagged Unidentified Species in the Ouachita Highlands." *North American Wood Ape Conservancy*, Jan. 2017, p 8. https://woodape.org/images/stories/file/Tag%207%20 Two%20Appendices%20ver2.pdf.

87 Ibid., 10.

88 Harris, S., Cresswell, W. J., Forde, P. G., Trewhella, W. J., Woollard, T., & Wray, S. (1990). "Home-range analysis using radio-tracking data--a review of problems and techniques particularly as applied to the study of mammals." *Mammal Rev., 20*(2/3), 97-123. Department of Zoology, University of Bristol, Bristol, U.K.

89 Walter, W. D., Onorato, D. P., & Fischer, J. W. (2015). "Is there a single best estimator? Selection of home range estimators using area-under-the-curve." *Movement Ecology, 3*(1), 10. Retrieved from http://doi.org/10.1186/s40462-015-0039-4.

90 Bowman, Paul, et al. "Tracking a Self-Tagged Unidentified Species in the Ouachita Highlands." *North American Wood Ape Conservancy*, 10 Jan. 2017, pp. 24-25. https://woodape.org/images/stories/file/Tag%207%20 Two%20Appendices%20ver2.pdf.

91 Lyda, S. B., Hellgren, E. C., & Leslie, D. M. (2007). "Diurnal habitat selection and home-range size of female black bears in the Ouachita

Mountains of Oklahoma." *Proceedings of the Oklahoma Academy of Science, 87,* 55-64.

92 Smith, T., & Pelton, M. (1990). "Home Ranges and Movements of Black Bears in a Bottomland Hardwood Forest in Arkansas." *Bears: Their Biology and Management, 8,* 213- 218. http://www.jstor.org/sta-ble/3872921.

93 Godfrey, E. (2014). "Outdoors notebook: Wildlife Department confirms mountain lion sightings in Oklahoma." *The Oklahoman.* https://www.oklahoman.com/story/sports/columns/2014/11/29/outdoors-note-book-wildlife-department-confirms-mountain-lion-sightings-in-oklaho-ma/60781571007/.

94 *Bats of Oklahoma Field Guide.* Oklahoma Department of Wildlife Conservation (2013). https://www.wildlifedepartment.com/magazine/batfieldguide/index.html#p=2

Chapter 24

95 "DNA: Comparing Humans and Chimps." *American Museum of Natural History,* https://www.amnh.org/exhibitions/permanent/human-origins/understanding-our-past/dna-comparing-humans-and-chimps.

Chapter 25

96 Lameira, A. R. "Orangutan (*Pongo* spp.) whistling and implications for the emergence of an open-ended call repertoire: A replication and extension." *The Journal of the Acoustical Society of America.* https://asa.scitation.org/doi/10.1121/1.4817929.

Index

Media

Places

Universities, Museums, and Government Agencies

ABOUT THE AUTHOR

Michael Mayes is a Central Texas-based teacher of History and a former coach who has had a lifelong interest in mysteries of the natural world. He has been a member of the North American Wood Ape Conservancy since 2005 and currently serves as the Chairman of the group's Board of Directors. He has appeared on numerous internet podcasts and radio programs, including *Expanded Perspectives, Coast-to-Coast AM, The Paranormal Podcast, Beyond Reality Radio, Bigfoot & Beyond*, and *Sasquatch Tracks* to discuss his research, as well as an episode of *The Lowe Files* on the A&E Network. Michael is the owner and writer of the *Texas Cryptid Hunter* blog and the author of the illustrated children's book, *Patty: A Sasquatch Story* and the non-fiction title, *Shadow Cats: The Black Panthers of North America.*

Lightning Source UK Ltd.
Milton Keynes UK
UKHW020654050722
405403UK00010B/768

9 781949 501223